Nottingham Trent University
CLIFTON LIBRARY
www.ntu.ac.uk/llr

Enquiries: 0115 848 2175

Higher Education Institutions in a Global Warming World

The Transition of Higher Education Institutions to a Low Carbon Economy

RIVER PUBLISHERS SERIES IN MANAGEMENT SCIENCES AND ENGINEERING

Series Editors

CAROLINA MACHADO
University of Minho
Portugal

J. PAULO DAVIM
University of Aveiro
Portugal

Indexing: All books published in this series are submitted to Thomson Reuters Book Citation Index (BkCI), CrossRef and to Google Scholar.

The "River Publishers Series in Management Sciences and Engineering" looks to publish high quality books on management sciences and engineering. Providing discussion and the exchange of information on principles, strategies, models, techniques, methodologies and applications of management sciences and engineering in the field of industry, commerce and services, it aims to communicate the latest developments and thinking on the management subject world-wide. It seeks to link management sciences and engineering disciplines to promote sustainable development, highlighting cultural and geographic diversity in studies of human resource management and engineering and uses that have a special impact on organizational communications, change processes and work practices, reflecting the diversity of societal and infrastructural conditions.

The main aim of this book series is to provide channel of communication to disseminate knowledge between academics/researchers and managers. This series can serve as a useful reference for academics, researchers, managers, engineers, and other professionals in related matters with management sciences and engineering.

Books published in the series include research monographs, edited volumes, handbooks and text books. The books provide professionals, researchers, educators, and advanced students in the field with an invaluable insight into the latest research and developments.

Topics covered in the series include, but are by no means restricted to the following:

- Human Resources Management
- Culture and Organisational Behaviour
- Higher Education for Sustainability
- SME Management
- Strategic Management
- Entrepreneurship and Business Strategy
- Interdisciplinary Management
- Management and Engineering Education
- Knowledge Management
- Operations Strategy and Planning
- Sustainable Management and Engineering
- Production and Industrial Engineering
- Materials and Manufacturing Processes
- Manufacturing Engineering
- Interdisciplinary Engineering

For a list of other books in this series, visit www.riverpublishers.com

Higher Education Institutions in a Global Warming World
The Transition of Higher Education Institutions to a Low Carbon Economy

Editors

Ulisses M. Azeiteiro

University of Aveiro
Portugal

Walter Leal Filho

Hamburg University of Applied Sciences
Germany

João Paulo Davim

University of Aveiro
Portugal

River Publishers

Published, sold and distributed by:
River Publishers
Alsbjergvej 10
9260 Gistrup
Denmark

River Publishers
Lange Geer 44
2611 PW Delft
The Netherlands

Tel.: +45369953197
www.riverpublishers.com

ISBN: 978-87-93609-20-4 (Hardback)
    978-87-93609-19-8 (Ebook)

Contents

Preface

This book aims to contribute to the global debate on Sustainability in Higher Education, in particular, to the transition of Higher Education Institutions (HEIs) to a low-carbon economy (LCE).

The transition of HEIs towards a LCE is aligned with the Paris Agreement, and with Sustainable Development Goal 13: Take urgent action to combat climate change and its impacts. It is also consistent with the European Commission's Climate strategies and targets and with the aims of the European Climate Change Programme.

Transitioning to a LCE represents one of the most significant and urgent challenges we are facing, and Universities have a critical role to play in fostering a low-carbon future, especially by developing innovative solutions. This book intends to be a contribution to this discussion about Sustainability in Higher Education, namely the transition of HEIs to a LCE.

A total of 9 double-blind peer-reviewed chapters from Europe (3), Asia (4), Africa (1), and Australia (1), cover different subjects related to the theme of this book (e.g., carbon management, climate change mitigation initiatives, curriculum innovation, and campus greening, University's Green programmes, low-carbon development-based curriculum, carbon footprint assessments, environmentally sustainable transportation policies for HEI, ranking for sustainable development (SD) in HEI and sustainability challenges).

The chapter from Mikémina Pilo and Boris Odilon Kounagbè Lokonon entitled "Higher Education Institutions and Carbon Management: Insights from Africa" provides evidence that HEIs can be helpful in reducing greenhouse gases emissions in Africa. This study calls for African universities structural reform to explicitly include in their agendas carbon management strategies taking as an example South Africa Universities.

Lina Erlandsson, Petra Molthan-Hill, Alexandra Arntsen, and Amanda Smith in the chapter "Combating Global Warming through the Estate and Curriculum – A Whole-Institution Commitment at Nottingham Trent University" describe innovative climate change mitigation initiatives happening at Nottingham Trent University (NTU) and reflect upon how academics are working with other members of staff within the University to enable students and staff to gain further understanding of ways to prevent global warming, on both an individual and institutional level. By highlighting the fact that everyone has a role to play in combating climate change, the university has been able to create an inclusive range of cross-disciplinary projects, focusing on finding solutions through innovation and collaboration. NTU has created opportunities for students and staff to explore their own role in the transition to a LCE.

Judy Rogers and Karolina Bartkowicz in the chapter "Linking Curriculum Innovation and Campus Greening in the Transition to a Low Carbon Economy: A Case Study" argue that mitigation involves more than campus greening initiatives and education about climate change and reduction of CO_2 emissions, it involves instead a complex process of encouraging behavioural change to address consumption patterns and waste while at the same time providing increasing opportunities for participation and innovation. This chapter highlights the importance of integrating campus greening strategies with hands-on experiential project-based learning and teaching initiatives for students and staff, as an approach to pursue climate change mitigation and support HEIs through the transition to a LCE. The chapter discusses, through a multistage green roof pilot project delivered at RMIT University, methods for harnessing opportunities and addressing barriers to green roof uptake. The project resulted in institutional learning and change leading to further investment and implementation of permanent green roof projects that will provide research and educational opportunities into the future. Project-based experiential learning it is argued here can, therefore, lead to multiple benefits for learners, for teachers and for institutions in the transition to a LCE.

In the chapter "Pathways to a Low Carbon Economy: the Evolving Role of the University of the Philippines" Mark Anthony M. Gamboa, Kristine F. Aspiras, and Annlouise Genevieve M. Castro assess the evolving role of the University of the Philippines, which is the only national university in the country, in the transition of HEIs to a LCE. As national university, various initiatives have been put in place as it is duty-bound to contribute to SD. Foremost is the University's Green UP program that aims to promulgate policies across the system, including green building standards, energy audits, replacement of equipment to improve energy efficiency, water and solid waste management, monitoring of building maintenance standards, environmental management, and other related strategic initiatives in relation with climate change, disaster risk reduction and sustainability.

In the chapter "Progressive Trends in Implementing Climate Change Courses in Higher Education Curriculum at Symbiosis International University, Pune, India" Prakash Rao and Yogesh Patil evaluate current trends in course curriculum development in line with Global climate negotiations as well as national climate priorities. The pedagogical approach uses assessments of sectoral Greenhouse Gas (GHG) emissions through class projects and real-time exercises using the Institute emissions as a live example. The courses are also tied to innovative market-based mechanisms and bring in recent industry technologies and regulatory-driven approaches such as the Perform, Achieve and Trade (PAT), Renewable Energy Certificate (REC) schemes, Central Electricity **Regulatory** Commission (CERC). Low-carbon development-based curriculum within courses in the management discipline through the direct linkages between global climate change impacts and energy development is seen as an enabler for students to demonstrate low-carbon activities through class room courses as well as in practice.

Prakash Rao, Saravan Krishnamurthy, and Vishal Pradhan in the chapter "Commuters' Carbon Footprints – A Sustainability Case Study from Symbiosis International University, India" describe the initiative of commuters Carbon Foot Print (CFP) assessment at Symbiosis International University, Pune, India. This baseline study is the foundation for carbon footprint assessments aimed to reduce the University's impacts on global warming. Currently, in India, new HEIs operate from multiple campus locations. Decision makers in Indian HEIs who intend to assist the transition of HEIs to a LCE stand to benefit from this research. Beginning with functional changes in commuting choices, taking urgent actions to combat climate change, could inspire pan-India sustainability policies development to reduce HEI CFPs. Widespread implementable LCE options aid integrated sustainability practices.

Vishal Pradhan, Saravan Krishnamurthy and Prakash Rao in the chapter "Determinants of Employees' Perceptions, Commuting Culture and Environmental Sustainability at Symbiosis International University, India" describe a study conducted to develop an understanding of socio-cultural rationalities and behaviour among employee commuters at Symbiosis International University (SIU) in Pune, an Indian metropolis.

This social study discusses Pune's metropolis commuting culture of employees. While attaining environmentally sustainable transportation policies for HEIs at its heart, this chapter addresses the need for a more integrated understanding of the mitigation challenges and makes rational recommendations for Indian urban HEIs. Employees' willingness and participation in outlined policies would realise the desired behavioural change towards meeting the low-carbon emission goals.

Ana Marta Aleixo, Susana Leal, Walter Leal Filho, Susna Mendes and Ulisses M. Azeiteiro in the chapter "Rankings and Sustainability in Portuguese Higher Education Institutions: A Descriptive Analysis" address the issue of ranking in HEIs in Portugal. This research was conducted using the Portuguese public HEIs websites and presents a critical review of HEIs rankings in Portugal. The links between rankings and institutional commitment, advanced sustainability or the promotion of a positive image are discussed. This work results in the preliminary discussion of a proposal for an alternative ranking for SD in HEIs. An alternative ranking would take the system and subsystem activities of HEIs into consideration, and would constitute a starting point for further discussion when it comes to the development of the ranking for sustainability in HEIs in responding to the issue of holistic and integrated sustainability in Portuguese public HEIs.

In the chapter "Living Labs for Education for SD in the Context of Higher Education. Identifying Triggers and Drivers of Development in the Portuguese Universities" Arminda do Paço and Ulisses M. Azeiteiro address the topic of Living Labs for education for SD and co-production as emerging strategies for universities to address sustainability challenges. This chapter provides an overview on the panorama of living labs in HEI. Authors present some successful cases, two from Europe (UK) and three from North America (USA and Canada). For the Portuguese HEI are identified the barriers, challenges and obstacles to instrument sustainable initiatives, as is the case of the living labs implementation. Proposals for embedding sustainability in organisational culture and practice are discussed.

Given the variety of research, this book offers a diverse thematic/disciplinary and geographic overview of some current research and projects/action projects in Sustainability in Higher Education, namely the transition of HEIs to a low-carbon economy. In addition, the chapters address some important challenges and future developments, also giving insights to the discussion around consequences at multiple spatial, temporal, education, and socio-political scales and the multiple dimensions of Sustainability in Higher Education and the transition of HEIs to a low-carbon economy practiced in an interdisciplinary dialogue.

We would like to take this opportunity to thank all authors who submitted their manuscripts for consideration of inclusion in this book. And since the peer review was a double-blind process, we also thank the reviewers who have taken time to provide timely feedback to the authors, thereby helping the authors to improve their manuscripts, and ultimately the quality of this book.

Ulisses Miranda Azeiteiro
Walter Leal Filho
João Paulo Davim

List of Contributors

Alexandra Arntsen, *Nottingham Business School/NTU Green Academy, Nottingham Trent University, 50 Shakespeare St, Nottingham NG1 4FQ, England*

Amanda Smith, *School of Animal, Rural and Environmental Sciences, Nottingham Trent University, Brackenhurst Campus, Southwell, NG25 0QF, England*

Ana Marta Aleixo, *Universidade Aberta, Portugal*

Annlouise Genevieve M. Castro, *Faculty Room 4, School of Urban and Regional Planning, University of the Philippines, Diliman, E. Jacinto St., UP Diliman Campus, Quezon City, Philippines*

Arminda do Paço, *Department of Business and Economics and NECE, University of Beira Interior, Covilhã, Portugal*

Boris Odilon Kounagbè Lokonon, *Centre de Recherche en Entreprenariat, Croissance et Innovation (CRECI) & Laboratoire de Recherche en Economie et Gestion (LAREG), Facultè des Sciences Economiques et de Gestion (FASEG)-Universitè de Parakou (UP)-Benin*

Judy Rogers, *School of Architecture and Design, RMIT University, Melbourne, VIC, Australia*

Karolina Bartkowicz, *School of Property, Construction and Project Management, RMIT University, Melbourne, VIC, Australia*

Kristine F. Aspiras, *Faculty Room 4, School of Urban and Regional Planning, University of the Philippines, Diliman, E. Jacinto St., UP Diliman Campus, Quezon City, Philippines*

Lina Erlandsson, *NTU Green Academy, Nottingham Trent University, 50 Shakespeare St, Nottingham NG1 4FQ, England*

Mark Anthony M. Gamboa, *Faculty Room 4, School of Urban and Regional Planning, University of the Philippines, Diliman, E. Jacinto St., UP Diliman Campus, Quezon City, Philippines*

Mikémina Pilo, *Faculté des Sciences Economiques et de Gestion (FASEG)-Université de Kara-Togo*

Petra Molthan-Hill, *Nottingham Business School/NTU Green Academy, Nottingham Trent University, 50 Shakespeare St, Nottingham NG1 4FQ, England*

Prakash Rao, *Department of Energy and Environment, Symbiosis Institute of International Business, Symbiosis International University, Pune, Maharashtra, India*

Saravan Krishnamurthy, *Symbiosis Centre for Information Technology, Symbiosis International University, Pune, Maharashtra, India*

Susana Leal, *Escola Superior de Gestão e Tecnologia, Instituto Politécnico de Santarém, and Life Quality Research Centre, Portugal*

Susana Mendes, *Escola Superior de Turismo e Tecnologia do Mar, Instituto Politécnico de Leiria, Portugal*

Ulisses M. Azeiteiro, *Department of Biology and Centre for Environmental and Marine Studies (CESAM), University of Aveiro, Aveiro, Portugal*

Vishal Pradhan, *Symbiosis Centre for Information Technology, Symbiosis International University, Pune, Maharashtra, India*

Walter Leal Filho, *Research and Transfer Centre Applications of Life Sciences, Hamburg University of Applied Sciences, Hamburg, Germany*

Yogesh Patil, *Symbiosis Centre for Research and Innovation, Symbiosis International University, Pune, Maharashtra, India*

List of Figures

List of Tables

List of Abbreviations

AR4	Fourth Assessment Report
CBO	Community Based Organization
CCCS	Cost, Convenience, Comfort and Safety
CDM	Clean Development Mechanism
CFP	Carbon footprint
CIRT	Central Institute of Road Transport
CMB	Common Method Bias
COP	Conference of Parties
COP	Conferences of the Parties
DRT	Demand Responsive Transport
EFA	Exploratory factor analysis
EMS	Environmental Management System
GHG	Green House Gas
GHG	Greenhouse Gases
HEA	Higher Education Academy
HEIs	Higher Education Institutions
ICT	Information and communication technology
iiE	Investors in the Environment
INDCs	Intended Nationally Determined Contributions
IPCC	Intergovernmental Panel on Climate Change
LCE	Low Carbon Economy
LiFE	Learning in Future Environments
LLCI	Lower limit of confidence interval
LUCF	Land-Use Change and Forestry
MLFA	Maximum Likelihood Factor Analysis
MoEF	The Ministry of Environment and Forest, Government of India
MOUD	The Ministry of Urban Development, Government of India
NAPCC	National Action Plan on Climate Change
NBS	Nottingham Business School
NTU	Nottingham Trent University

NUS	National Union of Students
PAT	Perform, Achieve and Trade
PECT	Peterborough Environment City Trust
PMC	Pune Municipal Corporation
PRME	Principles for Responsible Management Education
QAA	Quality Assurance Agency for Higher Education
REC	Renewable Energy Certificates
SD	Sustainable Development
SDA	Structure Decomposition Analysis
SDG	Sustainable Development Goal
SiP	Sustainability in Practice
SIU	Symbiosis International University
SME	Small and Medium Enterprises
SUTP	Sustainable Urban Transport Program
TILT	Trent Institute of Learning and Teaching
ULB	Urban Local Body
ULCI	Upper limit of confidence interval
UNEP	United Nations Environment Programme
UNFCCC	United Nations Framework Convention on Climate Change
UNISA	University of South Africa

1

Combating Climate Change Through the Estate and Curriculum – A Whole-Institution Commitment at Nottingham Trent University

Lina Erlandsson[1], Petra Molthan-Hill[2], Alexandra Arntsen[2] and Amanda Smith[3]

[1]NTU Green Academy, Nottingham Trent University, 50 Shakespeare St, Nottingham NG1 4FQ, England
[2]Nottingham Business School/NTU Green Academy, Nottingham Trent University, 50 Shakespeare St, Nottingham NG1 4FQ, England
[3]School of Animal, Rural and Environmental Sciences, Nottingham Trent University, Brackenhurst Campus, Southwell, NG25 0QF, England

Abstract

Nottingham Trent University (NTU) is at the forefront of sustainability in the Higher Education sector in the United Kingdom. Investment in climate change mitigation strategies plays a big role within estates and operations, with a number of important projects happening across NTU's four campuses. Projects such as 'Carbon Elephant' a carbon reduction programme and investments in renewable energy sources demonstrate a whole-institution commitment to combat anthropogenic climate change. This pledge is also reflected in the curriculum, with a range of innovative climate change education projects including: cross-disciplinary climate change games, an online module exploring sustainable energy and a greenhouse gas consultancy project for undergraduate business students. Recently, the University adopted a whole curriculum approach where every student has to address at least one of the United Nations' Sustainable Development Goals (SDGs), for instance SDG 13: Climate Action in their studies.

These projects, together with a range of other initiatives, highlight the importance of cross-disciplinary collaboration in addressing climate change related challenges and empowering behaviour change through education. By taking a unique approach to connecting projects within operations and estates with learning and teaching, NTU has created opportunities for students and staff to explore their own role in the transition to a low-carbon economy.

The aim of this chapter is to shine a light on innovative climate change mitigation initiatives happening at NTU and reflect upon how academics are working with other members of staff within the University to enable students and staff to gain further understanding of ways to prevent climate change, on both an individual and institutional level. By highlighting the fact that everyone has a role to play in combating climate change, the university has been able to create an inclusive range of cross-disciplinary projects, focusing on finding solutions through innovation and collaboration.

1.1 Introduction

Nottingham Trent University (NTU) is located in the East Midlands region of the United Kingdom. The university houses 28,500 students divided into eight academic schools and four different campuses. With around 640 different courses/programmes on offer, the university attracts a large number of international students as well as postgraduates. NTU is a teaching intensive, research active university and in 2015, the research excellence at the University was recognized when NTU was awarded the Queen's Anniversary Prize, a scheme recognizing world-class research and achievements in UK Further and Higher Education.

The University is dedicated to 'Creating the University of the Future', as stated in the 2015 strategic plan, with one of the main aims to "continue to be recognised as a leading exemplar of an environmentally responsible and sustainable organisation" (Nottingham Trent University, 2015). NTU is currently seen as one of the most sustainable universities in the UK with award-winning projects from both Estate and Curriculum. In 2016, NTU was again rated number 1 in the People and Planet University League, an independent league table of UK universities ranked by environmental and ethical performance in 13 different categories (People and Planet, 2016). The University also achieved, as the first Higher Education institution in the UK, a Gold Award in the Learning in Future Environment (LiFE) index (University Business, 2016), a scheme to improve social responsibility and environmental performance in UK Further and Higher Education providers (EAUC, 2017a).

NTU is dedicated to working towards fulfilling the United Nations Sustainable Development Goals (SDGs). The institution-wide commitment to the goals is seen in the curriculum development exercise *Curriculum Refresh* and the Business School at the University is a signatory of the United Nations Global Compact initiative PRME (Principles for Responsible Management). The projects described in this chapter especially contribute to SDG 13: Climate Action, SDG 7: Affordable and Clean Energy, and SDG 17: Partnership for the Goals (United Nations, 2017a).

The dedication to sustainability at NTU can also be seen through the appointment of two separate designated teams working on sustainability-related issues. The NTU Green Academy is responsible for embedding sustainability in teaching and learning and was initiated as part of the Higher Education Academy (HEA) Green Academy Change programme (Puntha et al., 2015). The Green Academy has developed from a temporary project to a permanent team working with stakeholders across both the University and the local community. The NTU Environment Team covers sustainability in estate and operations; from engaging students and staff with environmental issues across the four campuses, to diverting waste from landfill and ensuring environmental standards are met on all new building developments and refurbishments.

The combined work of these teams, together with the academic schools enables a whole-institution approach to combating climate change by reducing emissions from domestic activities and by providing our students with the right skills and knowledge to contribute to positive change in this area. Early on, NTU followed a Living Lab approach (Willats et al., 2017), seeking opportunities to connect the estate to the curriculum. In the following sections, different estate projects will be introduced first, followed by innovative teaching projects. We will then highlight some projects that link estate and curriculum, and finally reflect on the whole institution approach of NTU.

1.2 Climate Change in Estate and Operations

There is a commitment from the University to reduce negative environmental impacts, in particular to address climate change. The current Carbon Reduction Strategy for the estate is to have an absolute reduction of the university's carbon footprint of 48% by 2020/21 (Anderson & Brooks, 2010). The plan considers all energy consumption and carbon emissions from University operations: direct emissions (for example from gas heating on campus), indirect emissions (electricity consumption) and other indirect emissions (for example from supply chains and contractors). As stated in NTU's energy policy, the university is committed to optimizing energy procured from low carbon and renewable sources in order to reduce the carbon footprint of the University and its reliance on fossil fuels. (Nottingham Trent University, 2017a).

By implementing a number of projects and strategies to both limit the domestic carbon emissions and raise awareness of the issues and the power of individual action, the NTU Environment Team is leading the whole university in the important work towards the ambitious goals stated in University policies and in the Strategic Plan.

1.2.1 Carbon Elephant

Carbon Elephant is a University-wide initiative aiming to engage staff and students in the University's commitment to reduce its carbon emissions. Managed by the Estates department, the project plays a vital part in the University's carbon reduction policy, enabling all departments across the University to contribute towards the goals. Some of the key stakeholders driving the project forward include Estates and Resources, Internal Communications, Finance Management, Information Systems, the Students' Union, and Human Resources.

The project was launched in October 2012 as a competition to find the most energy efficient school or department at the University. Over 70 *carbon champions* signed up to help their department to start saving energy and top the monthly league table. The electricity use in each building is measured each month and compared against the building's baseline in the same period in the previous year. Any net savings based on reductions over the year are awarded to the departments to be re-invested in environmental projects of their choice.

The *Carbon Elephant* initiative engages the wider University staff to reduce the University's domestic carbon emissions by offering financial incentives based on the net energy savings. The project has also led to carbon reducing initiatives such as the installation of renewable technologies and energy efficiency measures in many of the University buildings. It has also contributed to a raised awareness of carbon reduction initiatives at the

University and measures that can be taken by individuals to reduce their own energy consumptions, all in line with the targets of SDG 7, Affordable and Clean Energy (United Nations, 2017b).

1.2.2 Green Impact

The behavioural change programme *Green Impact* is now in its 6th year at NTU, aiming to turn the University into a more sustainable workplace through engaging students and members of staff across the university.

Led by the National Union of Students (NUS) in the UK, the scheme aims to bring together students and staff to make positive changes on campus, in the curriculum and in the local community (National Union of Students, 2017). Staff and students at the University work in departmental teams to fulfill a wide range of criteria relating to sustainability and social responsibility. NTU students audit the teams, assessing the work the team has undertaken during the year.

The *Green Impact* scheme has led to a number of positive practical initiatives around the university, such as energy saving measures in classrooms and laboratories. It has also contributed to positive behavioural change within our staff and student body and promoted a collaborative approach in the spirit of SDG 17, Partnership for the Goals (United Nations, 2017c).

1.2.3 Environmental Management System – EcoCampus

The whole-institution commitment can also be seen in the use of the Environmental Management System (EMS) EcoCampus, run by the Estates Team. This system monitors all areas and aspects of the university to make sure it lives up to a certain standard, from domestic activities to external contractors.

The EcoCampus scheme is a national EMS available for all Higher Education Institutions in the UK and beyond. The scheme highlights the achievements of institutions in key areas such as environmental sustainability and carbon reduction (EcoCampus, 2017).

The use of the EMS is an excellent example of NTU's whole-institution approach to combating climate change. Assessing and monitoring every aspect of the University enables it to set up ambitious goals for the future and engage all academic schools and departments in working towards those set goals. The University first signed up to the pilot cohort in 2005 and has received the Platinum Award for all campuses, indicating a fully operational, auditable EMS (Nottingham Trent University, 2017b).

1.3 Climate Change in Teaching and Learning

Education can be the mobiliser needed to create momentum and change. By embedding climate change education in teaching and learning, students will be prepared to take on an active role through both mitigation and adaptation initiatives and limit the impact of climate change on societies across the globe. The empowerment of the individual leads to the behavioural change needed to put pressure on legislators, companies, and governments to take their responsibility and transform society for the better (CarboSchools Consortium, 2010).

As seen in Rowson and Corner (2015) all disciplines have a role to play in combating climate change and contributing to positive change. It is therefore vital to embed these themes across all courses and disciplines, not just in the areas where climate change education traditionally is found.

The first projects stated below are simple ways of introducing climate change education in both the formal and informal curriculum of higher education institutions or further education institutions that does not require a large amount of financial resources or time. The latter example, the sustainability in practice certificate, will require resources in the form of financial and structural input from the institution. The whole institution approach described at the end of this section requires dedicated commitments from the senior management team. More information on how to obtain the teaching materials described will be given at the end of this chapter.

1.3.1 Climate Change Games

The pressing need to address climate change via cross-disciplinary partnerships is well-established (Makrakis et al., 2012); as is the urgent need to achieve effective mitigation and adaptation at multiple scales (Anderson, 2012).

To address these priorities the Trent Institute of Learning and Teaching (TILT) created a series of innovative learning initiatives under the title of 'Climate Games', namely: a *COP Negotiation Workshop* and the card game *Carbon Crush-The Nations*. The aim of these initiatives is to provide students (or staff, as the 'games' can be used for staff development purposes) with unique opportunities to experience forms of participation, negotiations and outcomes of an International, UN led conference addressing Climate Change – the Conference of Parties (COP) in Paris 2015. Furthermore, the 'games' encourage critical reflection on personal and societal responses to climate change via feedback briefings or discussions at the end of each 'game'.

Students or staff from any discipline or department can participate in the workshop or play the card game.

The *COP Negotiation Workshop* runs for two hours and the participants become members of a state delegation, constituency or Non-Governmental Organisation attending COP21 in Paris. The delegation works together to put forward key objectives in terms of climate change and negotiates with others groups towards a legally binding agreement. As a result of participating in the workshop students (or staff) not only gain a fascinating insight into the processes involved but also develop a critical depth of understanding of climate change politics. In addition, students develop and practice key skills for employability such as teamwork, advanced communication (particularly negotiation), problem solving and information synthesis. Feedback from participating students suggests that the students recognize the skills gained when participating in the workshop and that their participation could be used as an example of negotiation skills, problem-solving or team work in future job interviews.

The game *Carbon Crush-The Nations* (see Figure 1.1) is based on the popular card game of 'Top Trumps' and students (or staff) use their dealt cards to pitch countries against each other in terms of their carbon

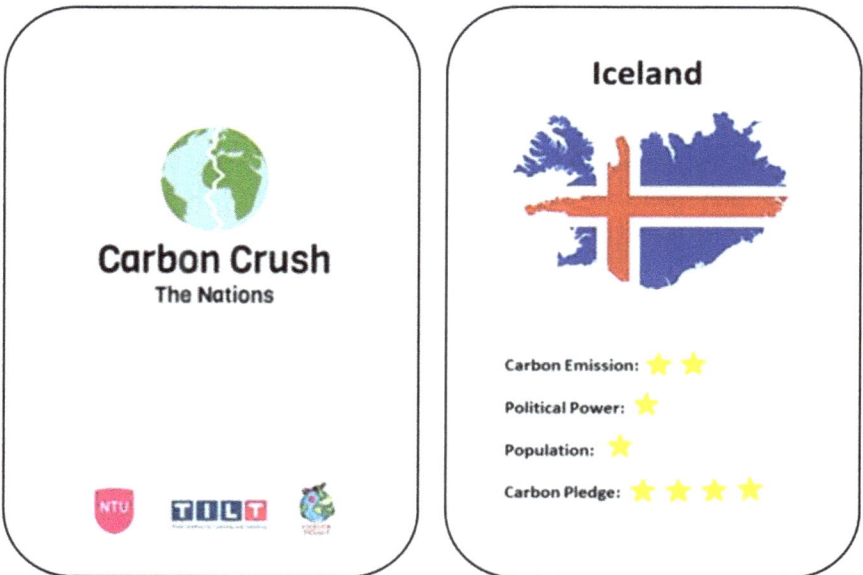

Figure 1.1   Examples of Carbon Crush: The Nations playing card.

emissions, carbon pledges, population and political power. This data is derived from the negotiations in Paris and similar to the workshop, the process of playing the game exposes geopolitical differences and inequalities in efforts to address climate change. It has been observed that during play the game is often halted whilst participants discuss the relative strengths or weaknesses of certain countries and the possible explanations associated with these. Furthermore, at the end of the game participants are asked to address a series of questions that encourage critical analysis of the key issues including:

- Which countries are the biggest emitters of carbon?
- Which have the best carbon pledges?
- How are populations, emissions and carbon pledges related?
- How do you think political power impacts on global carbon pledges?
- How could less powerful actors increase their power/influence?

1.3.2 The Greenhouse Gas Management Project

The Greenhouse Gas Management Project was initiated in 2011 in Nottingham Business School (NBS), the Business Faculty within NTU. Since then approx. 600 students have helped over 150 companies and organizations to develop a Greenhouse Gas Management Plan for their organisation and to work actively towards reducing Greenhouse Gas Emissions. In the last three years, NBS has teamed up with the local franchise of the nationwide not-for-profit accreditation scheme: Investors in the Environment (iiE) run by Peterborough Environment City Trust (PECT) (2017). The students act as consultants to the organizations advising and preparing them how they can achieve the iiE accreditation. For this innovative project, NBS and its partner won the Guardian University Award in Business Partnership 2015.

The project was at first part of an option and undertaken as a desk-based activity, however the ideas of the students were so innovative that the module leader wanted to share them with real organizations; and therefore initiated this project within the core curriculum for business undergraduates. Since then, students have not only developed sound environmental management systems for the participating organizations, but also developed unique recommendations for the companies; these range from a carbon-neutral Christmas decoration in a big modern shopping centre to carbon engagement events for Newstead Abbey, home to the poet Lord Byron and a historic house (in parts medieval) with its own challenges to reduce its carbon emissions.

A small research project undertaken in December 2016 calculated that the total recommended greenhouse gas emissions savings from the last two years of the project are 507, 435 kg CO_2e, averaging over 10 tons per organisation and 2 tons per student.

The project does not only reduce the carbon emissions of the participating organizations and contributes to SDG 13, Climate Action, but also increases the sustainability literacy of the students in their core curriculum. The project is fully described in a chapter by Molthan-Hill et al. (2017) highlighting the strong impact it has on employability skills and prospects. Some of the teaching methods used in this project have been integrated into a chapter by Dharmasasmita et al. (2017b) offering three fully developed seminars ready to be embedded into any (business) curriculum.

During the first year of running this project several units within NTU took part such as one library; another example how estate and curriculum can be linked. There are future plans to include other disciplines into this project, such as architecture and product design as this project would benefit from a transdisciplinary approach.

1.3.3 Energy Sustainability in Practice Certificate

The *Sustainability in Practice* (SiP) *Certificate* is an optional online module available for all students and staff at NTU. The certificate started out in 2013 and has now grown into including three different themes (Food, Clothing, and Energy) with approximately 4000 students starting the certificate each year. Some courses have chosen to embed the certificate in the formal curriculum and these students will undertake an online assessment test that is then part of their final piece of coursework. The themes of the certificate refer to the SDGs throughout and are in many cases the students' first introduction to the goals and Agenda 2030. Around 300 students complete and receive the certificate each academic year and their accomplishments are recognized at the *SiP Awards Night*, a celebratory event bringing together students and staff that have engaged with the certificate over the year. The best projects are awarded with special recognition and prizes include vouchers for a local, sustainable restaurant (Dharmasasmita et al., 2017a).

In 2016, an Energy version of the SiP certificate was developed to engage new student cohorts and to raise awareness of issues relating to the topic. Sustainability and energy are not simple issues. However, it tends to be reduced and simplified to a question of fossil fuels versus renewable energy, which are complex issues and very hard to identify with, something that creates a lack of engagement (Corner & Clarke, 2017). Therefore, the overarching aim of the Energy version of the SiP Certificate is to engage the participants and enable them to understand their role in creating a more sustainable energy system.

The aim when designing the Energy version of the SiP certificate was to get the participants to start noticing how they use both the more traditional forms of energy, such as oil and electricity as well as more "abstract" energy forms, such as energy gained from food and kinetic energy produced by humans. By highlighting the different types of energy and the accompanying issues, students also start to think of solutions. These solutions can include everything from charging the phone using kinetic energy to discovering a new way to make solar panels with a less toxic production. The most important thing is to raise awareness and get students to reflect on their own role in creating solutions, and making the rest of society more aware of sustainable energy.

The certificate contains four different sections, followed by a final project focusing on creating one's own solutions. The sections have been designed to engage students with the topic from a personal, disciplinary and inter-disciplinary perspective (Molthan-Hill et al., 2015), as seen in Figure 1.2. Debates and discussions between students in online discussion forums are expected and encouraged throughout the certificate. The discussion forums enable collaborative learning between students of different disciplines and levels of study, which can lead to both improved student engagement and achievement (Horizon, 2017).

The first session is designed to raise the student's awareness about their own relation to energy and the role energy sustainability plays in their personal lives. The students reflect on both traditional dilemmas relating to energy such as '*Are you worried about rising energy costs?*' as well as more abstract topics, for example '*How can energy be portrayed in art?*' and have to relate their personal views and values to these questions. The goal is to get the students to think of energy as more than just the traditional images of a wall socket, ligh bulb or a battery and expand their ideas on energy to movement, light, food, and connections.

The second session focuses on the links between energy, sustainability and the students' own area of study. The aim here is to give the students a better understanding of the discourse within their own field and show the role they can play in creating a more sustainable energy system. Academics from all eight schools have contributed with resources relating to energy, sustainability and the academic field. Here it is crucial that every department in the university helps to ensure high-quality content for every type of student in this section.

	Topic	Aim	Activity Example
Session One:	Student experience of sustainability and food/energy/clothing	Engage students on a personal level	Personal Core Value Survey
Session Two:	Sustainability and food/energy/clothing in your discipline	Facilitate disciplinary understanding of sustainability	Exploring web resources within and across disciplines
Session Three:	Connections between disciplines; identifying challenges	Facilitate interdisciplinary understanding of sustainability	Working with other disciplines to fulfil one or more of the SDGs
Session Four:	Global and local solutions to sustainability challenges	Identify disciplinary / interdisciplinary and supradisciplinary solutions	Interactive global map of solutions
Final Project Piece – poster, mood board, video (3 mins max) or practical project			

Figure 1.2 Structure of the Sustainability in Practice certificate. Adapted from Dharmasasmita et al. (2017).

After raising the student's awareness on energy sustainability in their own discipline, session three is designed to demonstrate how subject areas are working together to address these issues. This session uses the SDGs to exemplify interdisciplinary work on sustainability and with a special focus on energy. The session provides the student with a background on the SDGs and several examples on how the different sub-targets of the goals required interdisciplinary actions to be achieved. This section highlights the importance of interdisciplinary collaboration to overcome issues and create solutions. Students are encouraged to explore fields other than their own to broaden their perspective and reach a new understanding of energy sustainability.

The fourth and final session aims to combine the knowledge gained from the first three sessions and help the students start thinking of possible solutions to the challenges of energy sustainability. This section focuses on solutions on the local, regional and global level and gives examples of good practice and innovative solutions around the globe. This session is designed to encourage the students to start thinking of their own solutions, something that they will focus on in their final project.

The final outcome of the certificate is to present a solution to a sustainability issue relating to energy. Many of the projects presented by students tend to focus on climate change as a global issue and solutions that can help combating climate change on both local and global level. These student projects are then assessed by academics from the eight academic schools at the University who will give them their expert opinion and feedback relevant to the students' field of study. The best projects from each year are reused as teaching materials for future cohorts. This is something that promotes collaborative, student-led learning approach where students are presenting their own solutions to the climate change dilemma.

In 2015, the first *Sustainability in Practice Challenge Day* was organized by the NTU Green Academy. This was advertised as a chance to complete the Sustainability in Practice certificate in one day and get some practical sustainability experience. With the introduction of the two new themes, the Challenge Day has stuck to its original structure, with theoretical work via the online sessions in the morning, a campus tour led by a member of the Environment Team highlighting some of the most prominent sustainability features on each campus, a practical session in the afternoon relating to the students chosen theme and finally a chance to complete and present a solution to a sustainability issue. For the Energy Challenge Day, the practical session consists either of the two-hour climate change negotiation workshop or of a session led by a researcher from the Product Design department, showing the students how to reduce their energy consumption by making smarter choices in their everyday life. The Challenge Days have received very positive feedback from participating students, with 98% of the participants saying they would recommend the event to a friend.

1.3.4 Embedding the Issues in Core Curriculum

NTU's dedication to climate change education can also be seen in the taught curriculum at the University. By embedding the Future Thinking concept (see Figure 1.3) in the university-wide curriculum development exercise *Curriculum Refresh* (Simmons et al., 2016) the University is dedicated to providing students with the right skills to be part of future solutions rather than problems. Combating climate change is a very important aspect of this and preparing students for a future economy, not dependant on fossil fuels is a vital aspect of the Future Thinking concept. By giving students opportunities to explore these themes both in the formal and the informal curriculum, they can develop necessary skills and attributes that will help them going through the transition to a low-carbon economy.

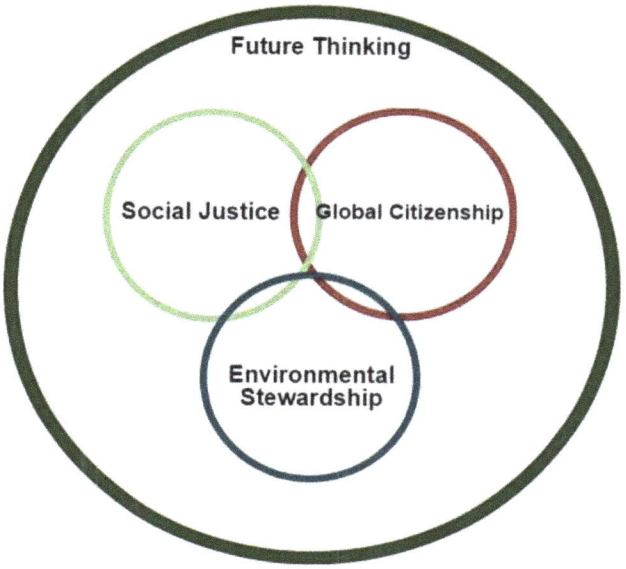

Figure 1.3 The Future-Thinking Model.

The Future Thinking concept is derived from the Quality Assurance Agency for Higher Education (QAA) and Higher Education Academy (HEA) report Education for Sustainable Development: Guidance for UK Higher Education providers (2014) and used to demonstrate that the Curriculum Refresh framework encompasses sustainability beyond themes or content to include knowledge and understanding, skills and attributes. It states that NTU is committed to sending students out into the world as global citizens, building on the links between sustainability literacy and employability. The framework has been selected to foster students' abilities to understand and contribute in meaningful ways towards current and future challenges in the area of sustainable development.

The *Curriculum Refresh* framework has also ensured that the United Nations' Sustainable Development Goals (SDGs) are embedded in all courses across the University. The rationale behind this is to clarify the sustainability concept by adopting the multifaceted set of goals that can bring attention to the various themes of sustainability and the opportunities for all sectors to contribute towards their fulfillment (Willats et al., 2017). The SDGs have proven to be an efficient tool to engage both students and staff, and are something that has been embedded in both staff development and student engagement activities.

1.4 Connecting Curriculum and Estate to Combat Climate Change

NTU wants to show that the University practices what it preaches. Therefore, it is vital to link back to the many innovative projects happening at our estate in teaching and learning. By making students more aware of the surrounding environment and the how their choices can affect greenhouse gas emissions and climate change in general, we can give students the opportunity to be change makers in the future. By connecting the estate and the curriculum, opportunities are created for students and staff to contribute to solutions to challenges the university currently faces.

1.4.1 University Wide Events

The Environment Team and the Green Academy at NTU often work together to set up events and happenings to shine a light on the sustainability work at the university, linking together estate and curricular projects. *Green Week* is an annual event with the aims of raising awareness of sustainability issues, both on a global and a local scale. Led by the Environment Team, *Green Week* engages a large number of both students and staff at the university each year and is always a much-appreciated event.

In December 2015, a *Climate Action Day* was organized to raise awareness of climate change and the COP21 negotiations in Paris. The day consisted of a wide range of activities such as a film screening, carbon friendly meals and stalls displaying ways to combat climate change. The event was led by the Green Academy in partnership with the Nottingham Trent Students' Union.

1.4.2 Using the Estate as a Teaching Resource

One way to inform students about how the University is trying to reduce its own impact on the climate are the *Campus Tours* offered to students participating in the SiP Challenge Days. These tours are led by a member of the NTU Environment Team and highlights specific features of our campuses. Some of the most appreciated features on these tours are the showcasing of the carbon negative Pavilion building and the investments in Solar PV installations generating approximately 350 Kw. These two projects raise students' awareness of the climate change mitigation measures taking place at the University, how NTU is contributing to the fulfillment of the SDGs and re-establishes the message communicated on the SiP Challenge Days. The Campus Tours are also an example of the importance of collaboration between the Environment Team and the Green Academy.

The *Estate and Community Case Studies* is another example of successful collaboration between a number of different teams and departments within the University. First initiated by the Trent Institute for Learning and Teaching (TILT) Education for Sustainable Futures grouping in 2016, the case studies display projects happening on campus and in the local community in an easily accessible format that enables these to be used as a teaching resource, as shown in Figure 1.4. To date, over 30 different projects have been offered to academics, varying from community volunteering to biodiversity projects on campus. Each of the case studies is linked to the SDGs and the resources are multidisciplinary and intended for use across a wide range of courses, throughout all eight schools at the University.

By making use of some of the features on campus available as teaching resources, both students and staff will get a first-hand experience of ways to combat climate change and that it is possible to contribute to positive change. One example of this is the *Foodshare Allotment*, located in the heart of the Clifton Campus. This is used as a volunteering opportunity for students to learn more about sustainable food production and the impacts of the current agricultural system. It is also used in formal teaching, with undergraduate students from two different courses (BA Primary Education and BSc Exercise, Nutrition and Health) working in the allotment as part of the formal curriculum. The aim of these sessions is to give students a better understanding of where the food they eat comes from, and the implications of their choices on both people and planet.

Moving forward, there are plans for developing the Living Labs concept (EAUC, 2017b) further at the University by embedding challenges and 'wicked' problems (Newman-Storen, 2014) in relation to the estate and the local community in the formal curriculum. One example of this is a project aiming to tackle the problem with packaging waste in the campus food outlets. This problem can be given to Product Design students that can focus on the design of the take away containers, but also to Marketing or Psychology students that can focus on promoting behaviour change. This will provide students with opportunities to contribute to the solution of real-life dilemmas and a chance to apply their knowledge and skills in the real world.

Community Case Studies as Learning Resources

Please complete form and email to Food4Thought@NTU.ac.uk. Any queries please contact the Green Academy on the same email address.

Which of the Sustainable Development Goals does this relate to?
(tick all that are relevant, or leave for Green Academy to complete if unsure)

1 NO POVERTY ☐	2 ZERO HUNGER ☐	3 GOOD HEALTH AND WELL-BEING ☐	4 QUALITY EDUCATION ☐	5 GENDER EQUALITY ☐	6 CLEAN WATER AND SANITATION ☐
7 AFFORDABLE AND CLEAN ENERGY ☐	8 DECENT WORK AND ECONOMIC GROWTH ☐	9 INDUSTRY, INNOVATION AND INFRASTRUCTURE ☐	10 REDUCED INEQUALITIES ☐	11 SUSTAINABLE CITIES AND COMMUNITIES ☐	12 RESPONSIBLE CONSUMPTION AND PRODUCTION ☐
13 CLIMATE ACTION ☐	14 LIFE BELOW WATER ☐	15 LIFE ON LAND ☐	16 PEACE, JUSTICE AND STRONG INSTITUTIONS ☐	17 PARTNERSHIPS FOR THE GOALS ☐	

Project Summary:

Project title:

Summary of activities:

Contact details for project:

Contact name:

Email address:

Especially suitable for: *(tick all that are relevant, or leave for Green Academy to complete if unsure)*

☐ School of Animal, Rural and Environmental Science
☐ School of Architecture Design and Built Environment
☐ School of Art and Design
☐ School of Art and Humanities
☐ School of Education

☐ Nottingham Business School
☐ Nottingham Law School
☐ School of Science and Technology
☐ School of Social Sciences

Teaching Ideas: *(to be completed by the Green Academy):*

Figure 1.4 Form designed by the Green Academy 2016 to capture case studies of community projects to be used as teaching materials.

1.5 The NTU Approach to Combating Climate Change

1.5.1 Connecting Curricula

NTU aims to give students more than just knowledge about sustainability issues and instead equip them with the right skills and attributes to be change makers and contribute to solutions to the challenges we are facing today. One of the many ways of providing students with opportunities to develop these skills is to present real-life examples on campus, showing that this is something we can do in practice rather than just in theory. By connecting innovative estate projects to the University's ambitious commitment to sustainability in the curriculum, this can become reality. Developing skills as well as knowledge plays an important part, both in climate change mitigation and adaptation strategies.

By re-establishing what is taught in the formal curriculum in the informal and the 'subliminal' curriculum, students are more likely to respond to the issues communicated (Tierney et al., 2015). This connection of curricula is vital for raising awareness and engaging both students and staff. Examples of this at NTU includes the Estate and Community case studies and the Campus Tours during SiP Challenge Days, projects that highlight measures taken on campus and within the local community, but also presents opportunities for individual students or members of staff to get involved and combat climate change.

1.5.2 Whole-Institution Approach

With two designated teams working on sustainability related themes, NTU possesses a unique starting point to cover all aspects of sustainability at the University. The collaborations between the two teams are vital to creating a vibrant sustainability community, bringing together and engaging both students and staff.

One aspect often highlighted is the importance of interdisciplinary collaboration and solutions, as previously seen in Anderson (2012), Mochizuki & Bryan (2015), Rowson & Corner (2015) and Makrakis et al. (2012). By offering University-wide projects and interventions, such as the climate change games, the Carbon Elephant project, and the Sustainability in Practice certificate, students and staff across the University can come together and work on finding solutions to the climate change dilemma. The interdisciplinary work is something the SiP Challenge Day students have really appreciated and stated as one of their favourite aspects of the day (Dharmasasmita et al., 2017). By encouraging interdisciplinary collaboration students and staff explore problems and solutions from different perspectives resulting in a wider understanding of the discourse around climate change and what role they can play in overcoming the issues.

The climate change games are also examples of the importance of individual academics as change agents as previously described in McRoy and Gibbs (2009). The dedication from engaged members of staff is of great significance here, creating opportunities to spread the message across the institution, inspiring more members of staff to get involved with the agenda. Encouraging already engaged academics and provide opportunities for them has therefore been vital in NTU's work, through projects such as the Sustainability Action Forum, a bi-monthly assembly bringing together colleagues from all academic schools to discuss sustainability in teaching and research, the TILT Education for Sustainable Futures group and the Carbon Elephant project, providing economic incentives to get involved with environmental projects and reduce one's own carbon footprint.

Another important factor in the whole-institution approach is leadership from the executive management, an essential factor in securing the staff and resource base to implement these projects. In order to create a coherent approach and a shared goal all stakeholders can work towards, leadership is vital to unite and create a coherent message around topics such as climate change as with any form of institutional change and commitment (McRoy & Gibbs, 2009).

1.5.3 Communicating a Coherent Message

There are many challenges linked to communicating climate change and engaging people with the topic (Corner & Clarke, 2017). There are many competing issues students and staff tend to be more likely focused on, partly because climate change can seem like a distant problem, which is not directly affecting the majority of our students and staff. The concept in itself is also quite complex, which contributes to a lack of interest and disengagement (Corner & Clarke, 2017).

Communications around climate change at NTU are therefore aiming to make the concept less abstract and show how the individual can make a difference to the global issue. Again, the collaboration between the Green Academy and the Environment Team is vital here in order to coordinate the communication of a coherent, strong message and to engage as many students and members of staff as possible.

The different stakeholders at the University ensure they communicate a coherent message around climate change mitigation to students and staff. In line with this communication, the University also wants to show that it practices what its staff preach and that the buy-in to these issues occurs on a whole-institution level, something that is shown by connecting the different curriculums.

1.6 Conclusion

The aim of this chapter has been to highlight the importance of collaboration between academics and professional services staff within Higher Education institutions to combat climate change, using the whole-institution approach at Nottingham Trent University as a practical example of how this can be implemented.

Higher Education institutions possess a unique opportunity to make a difference by influencing both students and members of staff to make changes in both their private and professional lives, contributing to positive change both in their current role and in the future. It is imperative that Higher Education Institutions maximize the skills and knowledge of all staff to inform, influence and inspire their students to take climate change seriously. By utilising their significant estates as a teaching and learning resource, Higher Education Institutions can also research and demonstrate meaningful, real world solutions to global problems.

For any further information on how to obtain the projects stated above or guidance on how to embed the initiatives described in this chapter, please contact the NTU Green Academy: greenacademy@ntu.ac.uk

The authors would like to extend their appreciation to Grant Anderson, Zoe Thomson and the NTU Environment Team, Dr. Roy Smith (School of Arts and Humanities, NTU), Dr. Mark Weinstein (School of Social Science, NTU) and all members of the TILT Education for Sustainable Futures group for their contributions to the projects stated above.

References

Anderson, A. (2012). Climate change education for mitigation and adaptation. *J. Educ. Sustain. Dev.* 6, 191–206.

Anderson, G., and Brooks, S. (2010). *Carbon Elephant Plan – Nottingham Trent University*. Available: https://www4.ntu.ac.uk/sustainability/docu ment_uploads/151278.pdf [last accessed September 22, 2017].

CarboSchools Consortium (2010). *Global Change: From Research to the Classroom.* Available at: http://www.carboeurope.org/education/3rd-booklet-single-reduced.pdf [accessed June 12, 2017].

Corner, A., and Clarke, J. (2017). *Talking Climate – From Research to Practice in Public Engagement.* London: Palgrave Macmillan.

Dharmasasmita, A., Puntha, H., Molthan-Hill, P. (2017a). Practical challenges and digital learning: getting the balance right for future-thinking. *Horizon* 25, 33–44.

Dharmasasmita, A., Kennedy, E., Puntha, H., and Holmes, R. (2017b). "Climate change and greenhouse gas management" in *The Business Student's Guide to Sustainable Management*, 2nd Edn., ed. P. Molthan-Hill (Sheffield: Greenleaf Press).

EAUC (2017a). *Learning in Future Environments.* Available at: http://www.eauc.org.uk/life/home [accessed May 23, 2017].

EAUC (2017b). *Living Labs – Opportunities, Benefits and Challenges of Different Models Globally.* Available at: http://www.eauc.org.uk/living_labs_opportunities_benefits_and_challeng [accessed June 12, 2017].

EcoCampus (2017). *About EcoCampus.* Available at: http://moodle.loreus.com/mod/page/view.php?id=1242 [accessed May 25, 2017].

Horizon (2017). *NMC Horizon Report > 2017 Higher Education Edition.* Available at: http://cdn.nmc.org/media/2017-nmc-horizon-report-he-EN.pdf [accessed May 23, 2017].

Makrakis, V., Larios, N., and Kaliantzi, G. (2012). ICT-enabled climate change education for sustainable development across the school curriculum. *J. Teach. Educ. Sustain.* 14, 54–72.

McRoy, A., and Gibbs, P. (2009). Leading change in higher education. *Educ. Manage. Adm. Leadersh.* 37, 687–704.

Mochizuki, Y., and Bryan, A. (2015). Climate change education in the context of education for sustainable development: rationale and principles. *J. Educ. Sustain. Dev.* 9, 4–26.

Molthan-Hill, P., Winfield, F., Hill, S., and Baddley, J. (2017). "Work based learning: Students solving sustainability challenges through strategic business partnerships," in *Redefining Success: Integrating the UN Global Compact into Management Education*, eds. P. Flynn, M. Gudiæ and T. Tan (Sheffield: Greenleaf Press).

Molthan-Hill, P., Dharmasasmita, A., and Winfield, F. (2015). "Academic freedom, bureaucracy and procedures: the challenge of curriculum development for sustainability," in *Challenges in Higher Education for*

Sustainability, eds. W. Leal Filho and J. P. Davim (Cham: Springer), 199–215.

National Union of Students (2017). *Green Impact – About.* Available at: https://sustainability.nus.org.uk/green-impact/about [accessed June 12, 2017].

Newman-Storen, R. (2014). Leadership in sustainability: creating an interface between creativity and leadership theory in dealing with "wicked problems". *Sustainability* 6, 5955–5967.

Nottingham Trent University (2017a). *Energy Policy.* Available at: http://www4.ntu.ac.uk/sustainability/document_uploads/193280.pdf [accessed May 24, 2017].

Nottingham Trent University (2017b) *What is EcoCampus?* Available at: https://www4.ntu.ac.uk/sustainability/ecocampus/ecocampus/index.html [accessed June 12, 2017].

Nottingham Trent University (2015). *Creating the University of the Future.* Available at: https://www4.ntu.ac.uk/strategy/ [accessed January 17, 2017].

People and Planet (2016). *People and Planet University League – First Class Universities.* Available at: https://peopleandplanet.org/university-league [accessed May 22, 2017].

Peterborough Environment City Trust (PECT) (2017). *Helping to Create a Cleaner, Greener, Healthier Peterborough.* Available at: http://www.pect.org.uk [accessed June 12, 2017].

Puntha, H., Molthan-Hill, P., Dharmasasmita, A., and Simmons, E. (2015). "Food for Thought: a university-wide approach to stimulate curricular and extra-curricular esd activity," in *Integrating Sustainability Thinking in Science and Engineering Curricula*, eds. W. Leal Filho, U. M. Azeiteiro, S. Caeiro and F. Alves (Cham: Springer), 31–48.

Quality Assurance Agency for Higher Education & Higher Education Academy (2014). *Education for Sustainable Development: Guidance for UK Higher Education Providers.* Available at: http://www.qaa.ac.uk/en/Publications/Documents/Education-sustainable-development-Guidance-June-14.pdf [accessed June 12, 2017].

Rowson, J. and Corner, A. (2015). *The Seven Dimensions of Climate Change: Introducing a New Way to Think, Talk and Act.* London: RSA Action and Research Centre.

Simmons, E. A., McNeil, J., and Lamb, S. (2016). *Curriculum Refresh: A Whole-Institution Approach to Reviewing the Curriculum, Trent Institute for Learning & Teaching.* Nottingham: Nottingham Trent University.

Tierney, A., Tweddell, H., and Willmore, C. (2015). Measuring education for sustainable development: experiences from the University of Bristol. *Int. J. Sustain. High. Educ.* 16, 507–522.

United Nations (2017a). *Sustainable Development Goals.* Available at: https://sustainabledevelopment.un.org/sdgs [accessed June 13, 2017].

United Nations (2017b). *Sustainable Development Goal 7.* Available at: https://sustainabledevelopment.un.org/sdg7 [accessed June 13, 2017].

United Nations (2017c) *Sustainable Development Goal 17.* Available at: https://sustainabledevelopment.un.org/sdg17 [accessed June 13, 2017].

University Business (2016). *Going Green at NTU.* Available at: http://university business.co.uk/Article/going-green-at-ntu [accessed May 22, 2017].

Willats, J., Erlandsson, L., Molthan-Hill, P., Dharmasasmita, A., and Simmons, E. (2017). "A university wide approach to integrating the sustainable development goals in the curriculum – a case study from the Nottingham Trent University green academy," in *Implementing Sustainability in the Curriculum of Universities: Teaching Approaches, Methods, Examples and Case Studies. 2017, World Sustainability Series*, eds. L. Filho et al., (Cham: Springer International Publishing).

2

Higher Education Institutions and Carbon Management: Insights from Africa

Mikémina Pilo[1] and Boris Odilon Kounagbè Lokonon[2]

[1]Faculté des Sciences Economiques et de Gestion (FASEG)-Université de Kara-Togo
[2]Centre de Recherche en Entreprenariat, Croissance et Innovation (CRECI) & Laboratoire de Recherche en Economie et Gestion (LAREG), Faculté des Sciences Economiques et de Gestion (FASEG)-Université de Parakou (UP)-Benin

Abstract

Environmental sustainability issues are closely linked to carbon footprint. Failure to ensure lower greenhouse gases emissions means that the world will be forced to continue facing severe climate change. Although Africa's emissions are currently relatively low compared to other regions, the continent has a critical role to play for a move toward a world without anthropogenic climate change. Thus, in order to support policy making decision we looked at, in this chapter, the indirect role higher educational attainment can play in the achievement of such objective. Using dynamic panel model, this chapter provides evidence that higher education institutions can be helpful in reducing greenhouse gases emissions in Africa. A result which is encouraging is given in the absence of explicit emissions management plan in most of these institutions. Consequently, in order to strengthen this positive trend, the study calls for African universities structural reform to explicitly include in their agendas carbon management strategies. The example of South Africa Universities can be of help.

2.1 Introduction

Climate change has invited itself in our daily life. Evidence brought from the Fourth Assessment Report (AR4) of the Intergovernmental Panel on Climate Change (IPCC) established that climate has globally changed and will continue changing in medium and long term (IPCC, 2007). As main sources of this change, greenhouse gases (GHG) emissions are pointed out. The emissions of GHGs are originated from diverse sources. In 2014, substantial amount of human-induced GHG emissions were coming mainly from fuel combustion and fugitive emissions from fuel (55.1%), transport including international aviation (23.2%), industrial processes and product use (8.5%), agriculture (9,9%) and waste management (3.3%) (IPCC, 2014). In Africa, GHG emissions were from three main sources that are (i) anthropogenic fossil fuel, (ii) agriculture and (iii) land-use change and forestry (LUCF) as indicated in Table 2.1.

The impacts of climate change are widely documented in developed countries and developing countries (Curriero et al., 2002; Dell et al., 2009; Barrios et al., 2010; Jones and Olken, 2010; Dell et al., 2012; Di Falco et al., 2012; Hasegawa et al., 2016). Climate change is found to lead to negative consequences on economic activities across countries, and will continue to affect them if nothing is done in terms of adaptation and mitigation. Some positive climate change effects are expected in some parts of the world. To avoid dramatic impact of anthropogenic climate change, the international community has agreed to maintain global warming below 2°C compared to the pre-industrial period temperature. Translated into today's temperature, this means that actual temperature should not increase more than 1.2°C. To stay within this threshold, the scientific evidence reveals that the world must stop the growth in GHG emissions by 2020 at the latest, reduce them by at

Table 2.1   GHG emission sources in African Regions

Regions	Fossil Fuel Emissions	Per Capita Emissions (Mg CO_2-eq)	Agriculture	LUCF	Removals	Net Emissions
North Africa	0.344	2.1	0.120	0.051	0.038	0.476
East Africa	0.157	0.8	0.279	0.645	0.426	0.219
West Africa	0.289	1.3	0.307	0.386	0.513	0.469
Central Africa	0.029	0.4	0.067	0.450	1.181	−0.626
Southern Africa	0.432	8.5	0.055	0.011	0.081	0.417
Total	1.250		0.828	1.115	2.238	0.954

Source: Authors based on Valentini et al. (2014).

least half of 1990 level by the middle of the century and continue cutting them thereafter (IPCC, 2007). This calls for urgent and practical policies implementation.

In the move towards lower level of GHG emissions, developed countries are thought to play the central role because of their higher current emissions level. Indeed, industrialized countries are the main responsible of the emissions of GHG and consequently of anthropogenic climate change which affects every country of the world. However, although Africa's GHG emissions are currently relatively low compare to other regions, the continent has a critical role to play for a move toward a world without anthropogenic climate change. Africa cannot stay behind efforts to be made in terms of reducing GHG emissions, and also has to follow a cleaner path of development instead of developing while polluting. The failure of the continent to shift to a sustainable GHG management strategy will cancel the positive benefits from considerable efforts made by the rest of the world to cut its emissions. Consequently, African continent should be brought on board in the search for means to limit climate change to an acceptable level. A wide range of strategies offer the possibility of reducing GHG emissions. These strategies include technologies and practices for end-use energy efficiency in buildings, transport and manufacturing industries, renewable energy use, and clean energy.

Higher education institutions are expected to considerably influence GHG footprint directly and indirectly. Directly, higher education institutions are expected to implement GHG best management practices to reduce their emissions. These practices include drawing GHG baselines, the identification of options to reduce GHG emissions. Some higher education institutions in Africa such as South African universities have defined their agenda for low carbon development. Indirectly, higher education institutions are expected to affect the use of best management practices by consumers and producers. However, the empirical evidence of this role is needed. What role do higher institutions play in carbon management? More specifically, do highly educated people acquire certain externalities that increase their capability to implement emissions reduction practices? To answer this question one has to look at the effect of higher educational attainment on GHGs emissions.

The remainder of the chapter is organized as follows. Section 2.2 presents a synthetic literature review and Section 2.3 the conceptual framework. The materials and methods are reported in Section 2.4. The estimation results as well as their discussion are presented in Section 2.5, and Section 2.6 is relative to the conclusion.

2.2 Literature Review

Owing to the implications of climate change on economic activities, and natural systems among others across the world, emphasis is put on efforts towards the reduction of GHG emissions. The 21st Conferences of the Parties (COP 21) of the United Nations Framework Convention on Climate Change (UNFCCC) pledges to limit the increase in temperature to below 2°C and even to tend towards 1.5°C, and consequently advocates for cut back in GHG emissions. Moreover, the COP 21 advocates for the increases in investments in renewable energy technologies (IEA, 2015). Higher education institutions have a key potential role to play in the limitations of GHG emissions. Higher education institutions carry out research, train people and can improve their awareness on the harmful effects of climate change on economic activities as well as on natural systems among others and on carbon management techniques. Those institutions can also serve as examples by implementing themselves carbon management options to motivate the population to do so.

The importance of Higher education in addressing climate change is increasingly recognized (Nhamo and Ntombela, 2014). Rauch and Newman (2009) stated that according to President Levin from Yale University, universities are considered as a hub of scientists, and a natural place in terms of devising innovative strategies to reduce carbon emissions. For Button (2009), higher education institutions have a moral responsibility to address the challenge of reducing carbon emissions, through teaching practice, strategies, research and their own practical actions. Altan (2010) conducted a survey among UK higher education institutions to explore the efficacy of some internal intervention strategies relative to carbon reduction, such as technical, non-technical and management interventions. From the consultation responses, he found that there are a relatively high percentage of institutions (83%) that have embarked on both technical and non-technical initiatives. Sinha et al. (2010) documented GHG emissions from U.S. institutions of higher education. They found that average annual emissions from all institutional classifications were 52,434 metric tons carbon dioxide equivalent (MTCO2E), with emissions from purchased electricity, stationary combustion, and commuting accounting for approximately 80% of total emissions. It should be noted that according to them, in 2005, U.S. institutions of higher education emitted approximately 121 million MTCO2E, representing nearly 2% of total annual U.S. GHG emissions.

The determinants of GHG emissions have been investigated in the literature. For instance, Hu et al. (2017) explored factors behind change in aggregate GHG emissions in Chongqing, including differences at the sector level using a structure decomposition analysis (SDA) method to quantify sources of emissions growth. Their findings suggest that factors related to intensities and input-output structure were those that are crucial drivers of GHG emissions reduction, and that the main driver of emissions growth was increasing final demand. Nässén (2014) investigated the determinants of GHG emissions from Swedish private consumption using both time series and cross-sectional data. The findings indicate among others that the level of education showed a positive relationship with emissions.

Apart from the level of education, other factors are found to influence GHG emissions. Expenditures and dwelling type were found as important determinants of GHG emissions (Nässén, 2014). Carattini et al. (2015) found a non-negligible impact of trust on GHG emissions, supporting Ostrom's intuition on the social roots of pro-environmental behavior (Ostrom, 2009), using a dataset of 29 European countries over the period 1990–2007. This finding suggests that there are unconventional determinants of GHG emissions. The other determinants of per capita GHG emissions are per capita gross domestic product (GDP), the aggregated industrial sector's share in the economy to account for the structural change, trade openness, and per capita energy consumption (Carattini et al., 2015). It should be noted that Carattini et al. (2015) do not account for the possibility of non-monotonic relationship between per capita GHG emissions and per capita GDP as advocated by the Environmental Kuznet Curve (EKC) literature probably due to the fact that their research is on European countries. Indeed, energy use and GHG emissions should increase as income rises in low-income countries, and as income levels increase, societies have the awareness and means to implement costly environmental schemes, leading to reductions in emissions (Stolyarova, 2012). This relationship indicates the decrease in the level of GHG emissions from a certain level of income. Marrero (2010) rejected the EKC hypothesis for 27 countries of the European Union (EU27). Urban population structure and dynamics can also determine GHGs' emissions' trend. According to Makido et al. (2012), less fragmented and compact cities emit less CO_2 per capita emissions. They also established that too dense settlements in mono-centric form may lead to greater per capita CO_2 emissions.

2.3 Conceptual Framework

As stated previously, institutions of higher education have a key role to play in the carbon management. Nhamo and Ntombela (2014) identified three channels through which higher education institutions can influence carbon management. For these authors, "Globally higher education institutions are believed to play a critical role not only through research, education and training but also through providing solutions for the impacts of climate change in their own context" (Nhamo and Ntombela (2014, p. 208). Thus, institutions of higher education can affect carbon management through i) research; ii) education and training; and iii) providing contextual solutions (Figure 2.1). For Brite Green (2016), universities have to consider factors such as sustainability impacts and policy, as they can pose physical and financial risks, shape decisions about estates strategy, as well as influence curriculum design and research focus.

Through research higher education institutions can discover new techniques of carbon management. Research also can provide information on the effectiveness of carbon management technique to address climate change. Research findings can be disseminated to enable the population to adopt those techniques. These institutions are supposed to carry out research on climate change challenges to provide solutions such as renewable energies technologies to be developed to reduce GHG emissions. Moreover, they can include aspects relative to carbon management in the curricula and educate and train new generation of people and also influence the ways of thinking of the population.

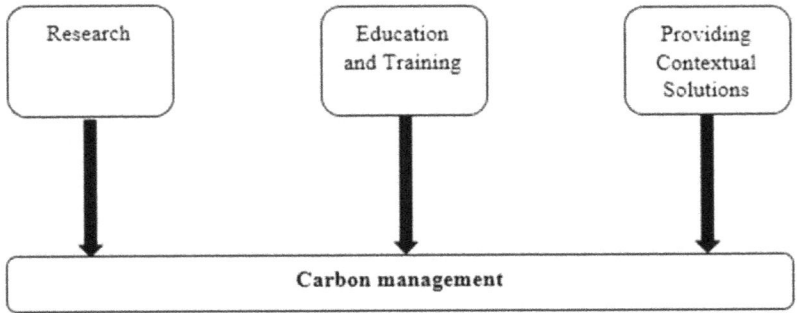

Figure 2.1 Carbon management channels of higher education institutions.
Source: Authors based on Nhamo and Ntombela (2014).

2.4 Materials and Methods

2.4.1 Empirical Model

The structure of our dataset gives us a considerable advantage. Indeed, one of the advantages of panel data is the fact that it allows better understanding of economics relationships dynamics. The main characteristic of the dynamic relationships is the presence of a lagged dependent variable among the regressors. GHG emissions dynamics can be modeled using similar dynamics. Thus, the level of GHG emission can be captured through the following relationship:

$$y_{it} = \delta y_{i,t-1} + x'_{i,t}\beta + \mu_{i,t} \tag{2.1}$$

Where δ is a scalar, $x'_{i,t}$ is $1 \times K$ regressors and β is a $K \times 1$ vector of parameters. In this formulation, $\mu_{i,t}$ is an error terms which we assume to follow one-way error component model.

$$\mu_{i,t} = \mu_i + v_{i,t} \tag{2.2}$$

With $v_{i,t} \sim IID(0, \sigma_v^2)$ and $\mu_i \sim IID(0, \sigma_v^2)$. The two components are independent from each other and among themselves. The dynamics described by the Equations (2.1) and (2.2) is characterized by two sources of persistence over time. An autocorrelation arising from the presence of a lagged dependent variable among the regressors and the individual effects characterizes the heterogeneity among the individuals (Baltagi, 2008). Even in the absence of serial correlation, the OLS estimator is biased and inconsistent (Sevestre and Trognon, 1985). Indeed, since y_{it} is a function of μ_i, it follows that $y_{i,t-1}$ is also a function of μ_i. Arellano and Bond (1991) proposed an efficient estimation technique known as General Method of Moment (GMM). The same authors argued that additional instruments can be obtained in dynamic panel data model. We consequently use Arellano and Bond technique to estimate GHG emissions dynamics. In that exercise, we used system GMM estimator since this estimator is shown to be powerful when weak correlation exist between the current and lagged values of model variables which is likely to happen in our analysis (Blundell et Bond, 1998). Two factors determined the variables included in the model: (i) the literature review on the topic and (ii) the availability of the data.

2.4.2 Data

In order to asses if individuals who have completed higher education level acquire certain externalities which enable them to implement GHG emissions reduction practices we gather data from two main sources: World Development Indicators (WDI) database (World Bank, 2016) of the World Bank and Barro and Lee (Barro and Lee, 2013) database on educational attainment. These are panel data covering the period 1990–2013. The data include 34 African countries, owing to data availability. The dependent variable is the GHG emission in CO_2 equivalent measured in metric ton per capita. Five variables are included in the vector of regressors. Higher educational attainment is the first regressor considered. This variable is measured as the proportion of population aged 25 and over that completed tertiary[1] level. This is our target variable. The second variable considered is access to clean fuels and technologies for cooking measured in % of population. The other independent variables are: expenses (% of GDP), urban population as % of total population, and agricultural machinery captured as the number of tractors per 100 square km of arable land.

2.5 Results and Discussion

We present descriptive statistics and discuss the results of the empirical estimations in this section. Table 2.2 presents the statistics of the variables. One can note from this table that the average GHGs emission per capita is estimated to 1.03 ton metric per year during the study period with a maximum of 10.26 recorded in South Africa. Regarding higher educational attainment, it has an average of 2.02 with a maximum of 8.1. This is to say that the percentage of the population aged 25 and over holding a diploma from a higher education institute in Africa varies from 0.75% to 8.10 %. The number of agricultural machinery for 100 square km also varies considerably within

Table 2.2   Descriptive statistics of the variables

Variables	Obs	Mean	Std. Err.	Max	Min
Higher education attainment	792	2.02	1.84	8.1	0.75
Greenhouse gases emissions	792	1.03	1.89	10.26	0.02
Access to clean energy	792	13.40	26.00	99.97	1.99
Urban population share	792	37.62	26.00	86.65	3.65
Expenses	792	9.91	11.85	43.84	0.17
Agricultural machinery	792	54.06	79.06	400.09	0.32

Source: Authors from World Bank (2016) and Barro and Lee (2013) databases.

---

[1]Tertiary education refers to education pursued beyond the high school level. This includes diplomas, undergraduate and graduate certificates, and associate's, bachelor's, master's and doctoral degrees.

the period of study and across the continent. It has a minimum of 0.32 recorded in Togo and a maximum of 400.09 recorded in Egypt. Access to clean energy is relatively low in the continent and depicts disparities across countries. Its mean is 13.40% with minimum and maximum amounting to 1.99% and 99.97%, respectively. Urban population share amounts on average to 37.62%, but its minimum is 3.65% while the maximum is 86.65%. As for expenses, the mean amounts to 9.91% of GDP, while the minimum is 0.17% and the maximum is 43.84%.

Before moving to the analysis of empirical results it is always interesting to have first a visual description of the data. For that end, a scatter plot of GHG emissions per capita as a function of higher educational attainment is presented (Figure 2.2). Observing this plot, one could refer to the points in the encircled area as potential outliers. However, these seem not to be. Indeed, the points in this area reflect higher level of emissions per capita. We consequently refer to these points as seemingly outliers since they cannot be taken as outliers. The idea of outliers being excluded from our mind, we can turn now to the trend line of the scatter plot. As one can see, there is likely a negative relationship between GHG emissions and higher educational attainment. As the educational attainment increases, GHG emissions decline.

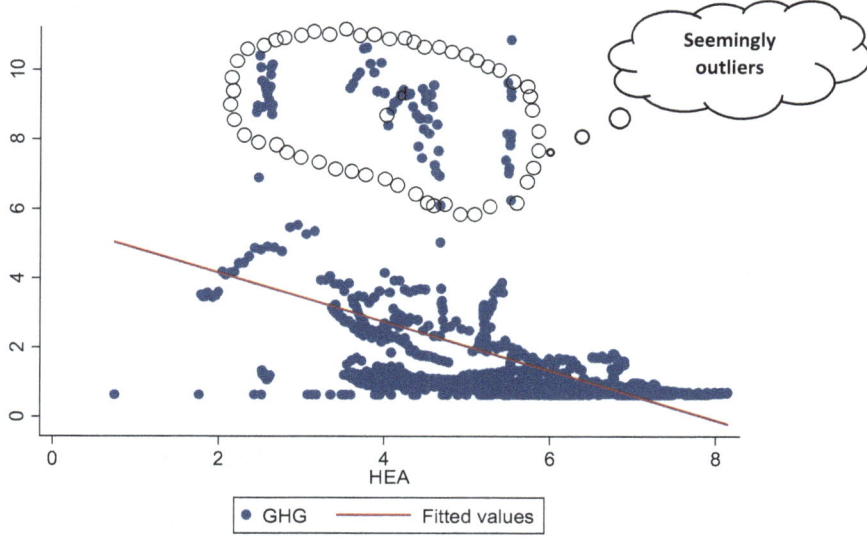

Figure 2.2   Relationships between GHG emissions and higher educational attainment.

The correlation coefficient computed seems to confirm this assumption. Indeed, the correlation coefficient indicates the existence of an inverse linear relationship between the two variables ($r = -0.32$). Thus, this preliminary finding suggests that higher educational attainment is important in reducing GHG emissions, even though the correlation is relatively weak. To see if this apparent relationship observed is confirmed, we now turn to the analysis of the empirical results from the dynamic panel data analysis.

Table 2.3 presents the estimation results from system dynamic panel-data. From the Sargan test we note that the instruments are valid (Prob > chi2 = 0.0000). This allows us to push further and learn the message that the estimates convey. The first message that can be drawn from Table 2.3 is that GHGs emissions per capita dynamics in Africa for specific period is firstly governed by the level of emissions of the previous period. Thus, 0.89 unit of a given period GHG emissions is explained by one unit of previous period emissions. This indicates technological rigidity of the continent. Indeed, the rate of technological change in Africa is very low and the move from one technology to another is a matter of decades. This is to say that emissions reduction in the short-term might be more complicated than stated in ongoing debates. Considerable time should be given to African countries to prepare their emissions reduction.

In this paper, we are particularly interested in the effect of higher education on the level of emissions. One can note from the estimates that higher education attainment can reduce GHGs emissions. Indeed, 1 unit achievement in higher education attainment reduces emissions by only 0.046 unit. That is to say that if as a result of educational policy the share of

Table 2.3 System dynamic panel-data estimation results. Greenhouses gases emissions per capita is the dependent variable

Independent Variables	Coef.	Std. Err	P > |z|
Greenhouses gas emissions lagged	0.8907***	0.0145	0.000
Higher education attainment	−0.0460*	0.0236	0.052
Agricultural machinery	0.0005	0.0004	0.240
Expense	0.0078***	0.0021	0.000
Share of urban population	0.0148***	0.0052	0.000
Access to clean energy	−0.0082***	0.0013	0.000
Constant	−0.3520 **	0.1742	0.053
Results of the sargan test of over identification	chi2(298) = 1741.56		
	Prob > chi2 = 0.0000		

Source: Authors estimates from World Bank (2016) and Barro and Lee (2013) databases.
Note: ***, **, * denote statistical significance at 1, 5 and 10 percent level respectively.

the population aged 25 and over with completed tertiary education level increases by 1 unit, the continent's emissions will fall by 0.04.6 unit. This highlights the important role higher education institutions play currently in the search for low GHGs emissions footprint. This result is so encouraging since explicit emissions management policies are taking time to emerge in such institutions on the continent. African universities have rarely developed agendas for low carbon development. It is common to see African universities continue using generators which function with fuel to supply electricity when there is black out. A good example can be learn from the University of South Africa (UNISA) which has embarked on climate/sustainability initiatives that show clearly the university's commitment to environmental issues (UNISA, 2012).

Let's now turn to the effects of other control variables on emissions. Access to clean technologies can strengthen emissions reduction strategies. It has, however, very limited effect on emission. This limited effect might be due to the difficulty in shifting from one technology to another and to the low effectiveness of the technologies classified as clean technologies on the continent. The emission reduction of these technologies needs to be revisited. Expenses are found to have a positive effect on GHG emissions. This finding suggests that the rise of expenses as percentage of GDP leads to increases in GHG emissions. Finally the share of urban population has a positive relationship with GHGs emissions. Indeed, the increase in urban population is associated with the rise in energy consumption, which is mostly from non-renewable sources leading to the increase in GHG emissions.

2.6 Conclusion

The world's commitment to maintain global warming at $2°C$ puts forth the search for practices and ways to manage GHGs' footprint. In the attempt to inform policies aiming to reduce emissions, we assessed in this chapter the possibility of relying on higher education institutions as emission reduction strategy. To be specific, through this chapter, we assessed if higher educational attainment provides spillovers effects that make citizens to more responsible regarding emission reduction focusing on the African continent. Because of the potential dynamic relationships that might exist between GHG emissions and the identified independent variable, we estimated dynamic panel data by the use of the system GMM technique since it has been shown to provide more consistent results compare to difference GMM.

From the findings we conclude that GHG emissions dynamics in Africa for a given period are mainly governed by past emissions. This is so more likely because of the technological rigidity of the continent. The implication of this is that African countries might need considerable time to cut down their emissions. Coming back to the focus of this chapter, the message from the estimations indicates the importance in reducing GHG emissions of higher education institutions. Indeed, we come to conclude from the results that 1 unit achievement in higher education attainment reduces emissions by 0.046 units. This wouldn't have been predicted given the absence of explicit agendas and policies in African universities. To strengthen this positive trend, universities of the continent should be committed to have explicit emission reduction agendas. Access to clean technologies appear also to be a good strategy in reducing emissions of the continent but its effect is lower than what should be expected and this question effectiveness of clean technologies in reducing GHG emissions at the current usage.

Appendix

List of the countries included in the analysis

No.	Name	No.	Name	No.	Name
1	Algeria	14	The Gambia	27	Sierra Leone
2	Angola	15	Ghana	28	South Africa
3	Benin	16	Kenya	29	Sudan
4	Botswana	17	Liberia	30	Swaziland
5	Burundi	18	Malawi	31	Tanzania
6	Cameroon	19	Mali	32	Togo
7	Central African Republic	20	Mauritania	33	Tunisia
8	Congo. Dem. Rep.	21	Morocco	34	Uganda
9	Congo. Rep.	22	Mozambique		
10	Cote d'Ivoire	23	Namibia		
11	Egypt. Arab Rep.	24	Niger		
12	Equatorial Guinea	25	Rwanda		
13	Gabon	26	Senegal		

References

Altan, H. (2010). Energy efficiency intervention in UK higher education institutions. *Energy Policy,* 38, 7722–7731.

Arellano, M., and Bond, S. (1991). Some tests of specification for panel data: Monte Carlo evidence and an application to employment equations. *Rev. Econ. Stud.* 58, 277–297.

Baltagi, B. H. (2008). *Econometric Analysis of Panel Data,* 4th Edn. West Sussex: John Wiley & Sons.

Barrios, S., Bertinelli, L., and Strobl, E. (2010). Trends in rainfall and economic growth in Africa: a neglected cause of the African growth tragedy. *Rev. Econ. Stat.* 92, 350–366.

Barro, R., and Lee, J.-W. (2013). A new data set of educational attainment in the world, 1950–2010. *J. Dev. Econ.* 104, 184–198.

Blundell, R., and Bond, S. (1998). Initial conditions and moment restrictions in dynamic panel data models. *J. Econom.* 87, 115–143.

Brite Green. (2016). *Carbon Management in the Higher Education Sector: A Guide to Good Practice.* London: BriteGreen Sustainable Strategy.

Button, C. E. (2009). Towards carbon neutrality and environmental sustainability at CCSU. *Int. J. Sustain. High. Educ.* 10, 279–286.

Carattini, S., Baranzini, A., and Roca, J. (2015). Unconventional determinants of greenhouse gas emissions: the role of trust. *Environ. Policy Gov.* 25, 243–257.

Curriero, F. C., Heiner, K. S., Samet, J. M., Zeger, S. L., Strug, L., and Patz, J. A. (2002). Temperature and mortality in 11 cities of the Eastern United States. *Am. J. Epidemiol.* 155, 80–87.

Dell, M., Jones, B. F., and Olken, B. A. (2009). Temperature and income: reconciling new cross-sectional and panel estimates. *Am. Econ. Rev.* 99, 198–204.

Dell, M., Jones, B. F., and Olken, B. A. (2012). Temperature shocks and economic growth: evidence from the last half century. *Am. Econ. J. Macroecon.* 4, 66–95.

Di Falco, S., Yesuf, M., Kohlin, G., and Ringler, C. (2012). Estimating the impact of climate change on agriculture in low-income countries: household level evidence from the Nile Basin, Ethiopia. *Environ. Resour. Econ.* 52, 457–478. doi:10.1007/s10640-011-9538-y

Hasegawa, T., Fujimori, S., Takahashi, K., Yokohata, T., and Masui, T. (2016). Economic implications of climate change impacts on human health through undernourishment. *Clim. Change* 136, 189–202.

Hu, Y., Yin, Z., Ma, J., Du, W., Liu, D., and Sun, L. (2017). Determinants of GHG emissions for a municipal economy: structural decomposition analysis of Chongqing. *Appl. Energy* 196, 162–169. doi: 10.1016/j.apenergy.2016.12.085

IEA (2015). *Energy and Climate Change: World Energy Outlook Special Report.* Paris: International Energy Agency.

IPCC (2007). *Climate Change 2007: Impacts, Adaptation and Vulnerability. Contribution of Working Group II to the Fourth Assessment Report of the Intergovernmental Panel on Climate Change.* Cambridge: Cambridge University Press.

IPCC (2014). *Climate Change 2014: Impacts, Adaptation, and Vulnerability. Part A: Global and Sectoral Aspects. Contribution of Working Group II to the Fifth Assessment Report of the Intergovernmental Panel on Climate Change*, eds C. B. Field, V. R. Barros, D. J. Dokken, K. J. Mach, M. D. Mastrandrea, T. E. Bilir, et al., Cambridge: Cambridge University Press.

Jones, B. F., and Olken, B. A. (2010). Climate shocks and exports. *Am. Econ. Rev.* 100, 454–459.

Makido, Y., Dhakal, S., and Yamagata, Y. (2012). Relationship between urban form and CO_2 emississions: evidence from fifty Japan cities. *Urban Clim.* 2, 55–67.

Marrero, G. A. (2010). Greenhouse gases emissions, growth and the energy mix in Europe. *Energy Econ.* 32, 1356–1363.

Nässén, J. (2014). Determinants of greenhouse gas emissions from Swedish private consumption: Time-series and cross-sectional analyses. *Energy* 66, 98–106.

Nhamo, G., and Ntombela, N. (2014). Higher education institutions and carbon management: cases from the UK and South Africa. *Probl. Perspect. Manage.* 12, 208–2017.

Ostrom, E. (2009). *A Polycentric Approach for Coping with Climate Change.* Washington, DC: The World Bank.

Rauch, J. N., and Newman, J. (2009). Institutionalizing a greenhouse gas emission reduction target at yale. *Int. J. Sustain. High. Educ.* 10, 390–400.

Sevestre, P., and Trognon, A. (1985). A note on autoregressive error component models. *J. Econom.* 28, 231–245.

Sinha, P., Schew, W. A., Sawant, A., Kolwaite, K. J., and Strode, S. A. (2010). Greenhouse gas emissions from U.S. institutions of higher education. *J. Air Waste Manage. Assoc.* 60, 568–573.

Stolyarova, E. (2012). *Carbon Dioxide Emissions, Economic Growth and Energy Mix: Empirical Evidence from 93 Countries.* Paris: Climate Economics Chair.

UNISA (2012). *Towards Environmental Sustainability: An Assessment of UNISA's Carbon Footprint and Appropriate Mitigation Actions.* Available at: http://www.unglobalcompact.org/system/attachments/21898/original/Appendix

Valentini, R., Arneth, A., Bombelli, A., Castaldi, S., Cazzola Gatti, R., Chevallier, F., et al., (2014). A full greenhouse gases budget of Africa: Synthesis, uncertainties and vulnerabilities. *Biosciences,* 11, 341–407.

World Bank. (2016). *World Development Indicators.* Washington, DC: The World Bank.

3

Commuters' Carbon Footprints: A Sustainability Case Study from Symbiosis International University, India

Prakash Rao[1], Saravan Krishnamurthy[2] and Vishal Pradhan[2]

[1]Symbiosis Institute of International Business, Symbiosis International University, Pune, Maharashtra, India
[2]Symbiosis Centre for Information Technology, Symbiosis International University, Pune, Maharashtra, India

Abstract

This chapter describes the initiative of commuters' carbon footprint (CFP) assessment at Symbiosis International University, Pune, India. A survey enabled comprehension of employees' commuting patterns to various campuses of the University. This baseline study is the foundation for CFP assessments aimed to reduce the University's impacts on global warming. Primary data yielded estimations of university Carbon emissions, commute choices of employees, per campus carbon footprints and eventually an assessment of the university's CFP in commuting aspects alone. These results indicate environmental co-benefits for each campus. The causality of CFP variation at campuses was construed from patterns of employee designations, campus locations, and vehicle types. Grouping of employees were deduced, with practical recommendations for each group and the University as a whole.

Currently, in India, new Higher Educational Institutions (HEI) operate from multiple campus locations. Decision makers in Indian HEIs who intend to assist the transition of HEIs to a low-carbon economy (LCE) stand to benefit from this research. Beginning with functional changes in commuting choices, taking urgent actions to combat climate change, could inspire pan-India sustainability policies development to reduce HEI CFPs. Widespread implementable LCE options aid integrated sustainability practices.

3.1 Introduction

Pune City, India is a hub of activity for educational institutions, IT companies and manufacturing industries, competing with other two million plus cities like Mumbai (Krishnamurthy et al., 2016; Dutt et al., 2016). High levels of urban development are predicted in Indian cities, in tandem with their low capacity for urban public infrastructure development and strong private property rights (Rode et al., 2014; Sellers et al., 2009). While tier two cities such as Surat or Hyderabad had initiated climate change action, Pune is yet to develop a major climate change action plan (Krishnamurthy et al., 2016; Sharma and Tomar, 2010). Bearing in mind that Pune and City of Bremen, Germany had joined hands as sister cities in 1976, it is appropriate to revisit the civil partnership. Noting previous failures in Pune such as building a tramway similar to Bremen, a multi-level climate governance in India is deemed essential. Well-defined efforts to synergize local and national level policies and further connect to global centers of information are critical (Beerman, 2017; Saroha, 2016; Jörgensen et al., 2015; Krishnamurthy et al., 2014).

With these new realizations, Pune was chosen to explore Higher Educational Institutions (HEIs) participation in low-carbon economy (LCE), specifically campus commuter carbon footprints (CFP) reductions. The importance of fast developing Asian cities and their impending emissions rises are noted (Shirgaokar, 2014; Pucher et al., 2005). A few studies within an Indian Metro City validate commuter's choices in reducing their footprint in a single organization, with empirical evidence extended to observe commuters' socioeconomic features and relevant travel choices (Paladugula and Rathi, 2013; Subbarao and Krishna Rao, 2013). The Indian cities selected in previous researches were either predominantly business centric (such as Mumbai and Bengaluru) or had recently exhibited rapid urbanization (such as Surat and Pune) (Shirgaokar, 2014; Fatima and Kumar, 2014; Subbarao and Krishna Rao, 2013; Paladugula and Rathi, 2013). Extensive studies aimed at healthy sustainability practices for Indian cities, had explored distinct aspects of public transport such as Bus Rapid Transit System (BRTS) or the commuters' energy attributes (Kathuria et al., 2016; Dash and Balachandra, 2016). However, the potential of HEIs in India has not been explored so far, especially concerning campus sustainability and commuter CFP assessment.

The research gap noted were baseline studies attempting to demonstrate the potential of educational organizations towards LCE. The current chapter tries to show the potential of HEIs' role here, selecting Pune, a popular education hub of India.

3.2 The Need for Carbon Footprint Assessment in HEIs of India

Symbiosis International University, Pune had formally initiated sustainability research studies in the year 2014, with a primary objective to explore and arrive at judicious goals for the educational community. Given the urgent needs of the urban India for modernizations in waste management, exploring waste management solutions was also set as an objective (Beermann et al., 2016; Mane and Hingane, 2012). Aligning to National Urban Renewal Mission (MOUD, 2017), critical research advises action by academic participants and improvements in essential services delivery such as transportation with equity (Kundu, 2014; Shen et al., 2011). In the process of goal setting aligned with national development, CFP estimates were adopted as a pivotal study, aiming at long-term neutralization of the university's carbon emissions. The baseline study thus developed is the theme of the current chapter. The research objective of this study is to estimate the Green House Gases (GHGs) level 1 emissions of Symbiosis International University (SIU) Pune campuses. To begin action as academic participants, isolatable illustrations of commuters' CFPs were considered as a useful scaffolding towards future policy making. The outcomes demonstrate a useful platform for further discussions and future strategies for Urban India, partially fulfilling the role of HEIs in sustainability practices of campuses.

Long-range national strategies need adaptation and mitigation agendas for India cities with rapidly expanding urban populations (Revi, 2008). Considering urban vulnerability to climate change, especially the poor, Revi (2008) advises a national adaptation program of action with multi-stakeholder engagement grounded in the institutional, sociocultural, and political realities of India. To develop leadership in sustainability practices for HEIs, proactively developing studies in carbon emissions is essential (Finlay and Massey, 2012). The research outcomes could affect positive influences on HEI policies for integrated planning and stimulate implementation of carbon friendly measures, suitable for urban universities (Finlay and Massey, 2012; Spirovski et al., 2012; Rao, 2011). Realizing the importance of carbon measurement as a critical step on the way to future planning, well-grounded current studies providing insights into corrective action are the need of the hour (Aroonsrimorakot et al., 2013; Doll et al., 2013; Revi, 2008).

Urbanization in Pune district had not reached sufficient planning benchmarks for livability (Jha, 2015). We compared Pune with other nations' cities, where sometimes a lack of urban population is considered economically unviable or where a new University became a new growth center for developments in the city. Pune's rapid expansion in urban population was concurrent with the establishment of many private academic institutions (Kantakumar, 2016; Nuzir and Dewancker, 2014; Pradoto, 2012). With a majority of the world's megacities predicted in Asia, Pune is expected to become one of the Indian megacities, with rapid economic development bolstering its rapid population expansion (Krishnamurthy, 2016; Jha, 2015). Noting lessons from developed countries, Universities in Asian cities need to become proactive in their vehicular carbon emission estimations. They need to develop scrupulous studies to confront urban issues in Tier 2 cities like Pune and develop new methods such as Information and communication technologies (ICT) to reduce commute to campus routines (Krishnamurthy, 2016; Clifford and Cooper, 2012; Roy et al., 2008).

3.2.1 HEIs in India Can Influence Transitions to LCE

Several India-centric studies have assessed sustainability practices. A particularly comparative study in sustainability indicated ecologically sophisticated responses of the local people, consequently qualifying them to the status of "ecosystem people" of their respective village areas (Dasmann, 1988; Duffield et al., 1998). Early studies on GHG mitigation in India focused on the agriculture sector, given its agro-based economy of India and the predominance of soil N_2O mitigation (Gupta et al., 2016; Garg et al., 2001). Prevalent glocalized applications of GHG emissions control research relate to rural India household cooking fuel. Shifting away from thermally inefficient biomass stoves was deliberated as a small mitigation step, but an implementation for common people of India to relate and consider larger GHG emissions control goals. These significantly persuasive implementations include a healthy transition towards non-polluting households for rural women with credible co-benefits of air pollution mortality avoidance (West et al., 2011; Venkataraman et al., 2005; Smith et al., 2000). India has a rich system of local knowledge applicable in many sustainability aspects, which is inordinately missing in urban environments (Sapkota et al., 2015;

Hansson and Wackernagel 1999). This temperament is keenly felt as a pattern of dis-embedding of modern urban societies, shifting its pressures on semi-urban and rural ecosystems with fading concerns to ecology, consequently placing a higher demand on urban citizens initiatives and multi-disciplinary-trans-sectors research interventions by academia (Rao and Leal Filho, 2015; Hansson and Wackernagel 1999).

Later studies shifted towards energy generation mechanisms with an emphasis on conversion to renewable energy (Schmid, 2012; Panwar et al., 2011; Ramachandra and Shruthi, 2007; Lewis and Wiser, 2007). Low-carbon technologies often compete and substitute among themselves (Shukla and Chaturvedi, 2012). While mitigation assessments abound among energy-intensive production sectors, very few research attempts have been made in non-energy-intensive service sectors such as education. Even though the best research capacities exist among academia, the HEI sub-sector has not adequately addressed the need for CFP assessments within their campuses. The current chapter aims to redress this lacuna and to inspire GHG mitigation implementations in many HEI campuses dispersed across the nation, a presumably insightful attempt for a nation-wide LCE adoption. The seeds of transitions need to be planted today (Yaduvanshi et al., 2016).

Compared to India, China faces more time pressure to transition towards LCE and conducts more research in the field (Messner et al., 2010). Currently, low-priced emission reductions such as commuters' carbon emissions in HEIs may seem insignificant compared to substantial emissions in sectors such as agriculture and manufacturing. In the early phases of climate change action, India's relatively low per capita emissions may not enthuse mitigation efforts. However, HEIs delaying mitigation opportunities today will increase dependence on other drivers of mitigation in the education sector (Saveyn et al., 2012; Shukla, 2012). Other carbon friendly drivers may include: policy instruments, local resistances, comprehensive technology adoptions such as electric vehicles replacements, predominating ICT over other instruction mediums to reduce commutes (Shukla, 2012; Roy et al., 2008). While nations may draw learning from other countries and non-HEI sectors, it is essential to develop low carbon solutions assimilated from one's national reality of HEIs, and evoke deep-seated values of native culture for mitigation (Mulugetta and Urban, 2010; Duffield et al., 1998). To develop HEI-specific learning, sustainability education across many university courses, developing participatory approaches to invite and engage stakeholder groups are essential for progressive integration into LCE (Aleixo et al., 2017: Disterheft et al., 2015).

In India, macro-measures such as subsidies for all consumers or proposed national emission reduction targets common to many sectors, diminishes a single-mindedness of purpose, especially among sub-sectors such as HEIs (Shukla, 2012; Saveyn et al., 2012). To summarize, in order to avoid catch-up decarbonizing efforts in later phases, India could wisely invest in sub-sector-wise GHG mitigation efforts, with Indian Universities at the research forefront and policy implementations (Saveyn et al., 2012; Messner et al., 2010; Mulugetta, 2010). To transition to LCE an essential step is establishing the need for implementable research prototypes that can pervade HEIs. To complement the sustainability paradigm, a commuter's survey and analysis portray CFP reduction potential within an Indian University.

3.3 Data Collection

An online survey was conducted yielding 355 useful responses, covering 14% of the university employee population. Respondent employees of SIU were from seven campus locations in the Pune district, Maharashtra state. Commuting behavior data were collected to develop a baseline study, as preliminary assessment of employee carbon emissions. The objectives of the survey were: to understand the University's potential impact of its high-carbon activities; to develop the foundation for an integrated study of sustainability practices through policy and operations. Analysis of primary data yielded key results regarding behavior patterns of commuters and per employee, per campus CFPs. The online questionnaire administered consisted of the following sections:

- Demography, commuting details (collected basic commuting towards campus details, sharing vehicles, peak hour traffic, and costs).
- Commuting Preferences (asked how commuters feel about commuting, gathered current commute choices, and considerations for change).

Descriptive statistics show initial visualizations and prominent relationships among the entities of study. Then, with selected variables concerning preferences, independence relationship between the first mode of commute and second mode of commute were analyzed with Chi-square method. It was found that first mode of commute is not independent of the selection of choice of the second mode of commute. The second Chi-square test showed that gender does not play a role in commuter decisions. After these exploratory steps, the suitable analysis identified was regression models. These regression models attempt to find the moderating effect of some variables (designation, gender, and location of campus) on the CFP of an employee.

3.3.1 Descriptive Statistics of Campus Commuters Survey

In Pune, there are seven campus locations of SIU. The campuses are located at Atur Centre, Khadki, Model Colony, S.B.Road (4 campuses close to Pune City Center), Hinjawadi, Viman Nagar and Lavale (3 in suburban Pune). SIU's main campus, Lavale is located farthest away from Pune city urban limits (in Mulshi Tehsil administrative sub-division of Pune district). The remaining campuses are within the expanded Pune city limits which had developed as business centers or educational hubs. Lavale campus provided maximum completed responses among all campuses. Nearly 50% responses were received from Non-teaching support staff and coordinators, whereas 49% were from administrators and teaching staff. Overall, the distance commuted has a marked difference in the variability between the types of campuses as operationally defined above. The responses received from male: female participants were in 59:41 ratios. Nearly 50% employees associated with Symbiosis institutes belong to 30–40 age group. 25% each are either younger are older than this age group. Respondents mentioned their first portion of commute (for instance, a personal vehicle) and the second portion of commute (for instance, public transport buses). 50% employees commute on two-wheelers, due to proximity of their residences to relevant campus locations and convenience. About 15% employees avail the facility of sym-biosis campus provided bus transport, which is significantly small. But those who commute using the second mode are opting campus bus facility, a total of 27%. Socio-demographic aspects of individual commuters were useful in profiling specific to distance or fuel consumed and were further explored in regression modeling. Increased vehicular population and sense of safety were noted as possible reasons for commuting by campus bus. 50% commuters use personal motorized vehicles to commute, of which 80% commute by two wheelers. Approximately 38% of this 50% share their vehicles with one to two colleagues on 2–3 days a week. A variability is observed in the cost spent per week on petrol.

3.4 CFP Computations

Carbon footprints were computed using the India GHG on-line calculator (India GHG Program, 2015). The on-line calculator uses India specific factors

Table 3.1   Carbon footprints of Employee Commuters, SIU, Pune

Designation	Sample Size	TOTAL CFP (Metric Tons)	Average Carbon Footprint (CFP) (Metric Tons)
Full-time faculty	124	101.72	0.82
Visiting faculty	3	2.39	0.797
Non-teaching (administrators, officers, etc.)	46	29.58	0.643
Non-teaching (coordinators, support staff, etc.)	182	134.71	0.765
Total	355	268.4	0.756

Source: Authors.

Table 3.2   Average CFP per capita each campus of SIU, Pune

Campus	Sample Size	TOTAL CFP (Metric Tons)	Average CFP (Metric Tons)
Lavale campus	132	128.1	0.97
S. B. Road campus	21	11.79	0.561
Atur Center campus	33	15.88	0.481
Viman Nagar Campus	28	18.54	0.662
Hinjewadi campus	68	59.94	0.881
Khadki campus	19	8.17	0.43
Model Colony campus	54	25.98	0.481
Total CFP	355	268.4	0.756

such as vehicle type for two-wheelers, diesel engine powered buses and India specific petrol consumption of small and medium size cars, enabled accuracy in computing CFPs. Summarized CFP values are tabulated below (Tables 3.1 and 3.2).

3.5 Data Analysis

3.5.1 Chi-Square Method

This research argues that commuting by mass transit (public or campus bus) and active mobility (walking or bicycling) are carbon-friendly modes. These two commonly endorsed options were analyzed. The dependence relation was found between the commuters taking any first mode and the endorsed second mode of transport (χ^2 = 35.87, p < 0.001 at 3 df). This encouraging finding uncovers potential for future studies on mode choice behavior among different attributes of commuters. To clarify whether gender and the aforementioned carbon friendly modes are dependent, the commuters based on their first mode of transport were segregated. Among the (first mode) 172 two-wheeler commuters, men and women are indifferent about opting carbon-friendly commuting options ($\chi^2 = 0.047$, $p = 0.83$ at 1 df). Also, amongst all of the 68 car commuters, men and women employees are indifferent about commuting by carbon friendly options ($\chi^2 = 1.515$, $p = 0.22$ at 1 df). Thus, gender does not play a meaningful role in the choice of commuting towards carbon friendly travel modes when controlled by overruled choices of mobility (car, two-wheeler). These results enable policy makers and researchers to initiate future research to elaborate on mode choice behavior among different attributes of commuters.

3.5.2 Regression Models

We then proposed several simple moderation models to assess their ability to predict Total CFP. Here, X is the predictor variable (X = Distance from home and time required for commuting), M is the moderator variable (M = major commuting mode type, gender, designation, and campus location), XM is the interaction variable, and CFP is the outcome variable. The moderation models are shown in Figure 3.1.

$$\text{CFP} = \beta_0 + \beta_1 X + \beta_2 M + \beta_3 XM + \varepsilon.$$

Overall, the moderation models were substantial ($p < 0.001$ and $R^2 > 0.5$), but occasionally, the interaction effect size between the focal predictor X and the moderator was modest to low. The distance from home and major commuting mode-type were significant predictors of total CFP. The interaction effect was non-significant ($p = 0.14$). This means as distance from home is increasing, the increased value of total CFP may not be due to the moderator,

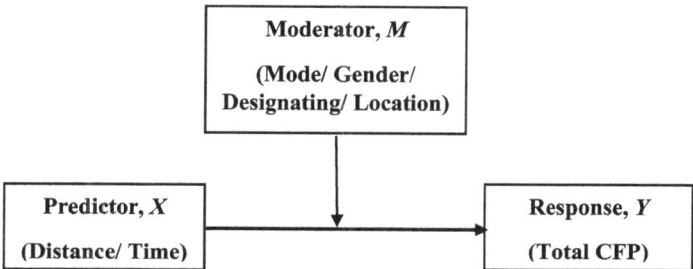

Figure 3.1   Consolidated conceptual diagram.

Source: Authors.

Please read diagram considering one Predictor and one Moderator at a time.

i.e., commuting mode. This argument is further countered by the significant conditional effect of distance from home on total CFP at various options of commuting mode (such as two-wheeler, car, and bus) (p-values < 0.05). Thus, it is confirmed that solo commuters in cars contribute to larger total CFP, as the distance to campus increases.

The time required for commuting and mode are significant predictors of total CFP. The interaction effect between time and mode was non-significant ($p = 0.25$). This means as the commute time increases, the increased value of total CFP may not be due to the commuting mode. This argument is further countered by a significant conditional effect of commute time on total CFP at various categories of the moderator, i.e., commuting mode. Thus, commuting mode has significant conditional impact on predicting total CFP (p-values < 0.05).

The distance from home is a significant predictor of total CFP ($p < 0.001$). Gender is a marginally non-significant predictor of total CFP ($p = 0.053$). The interaction effect is non-significant as well ($p = 0.28$). This means as commute distance increases, the increased value of total CFP may not be due to the moderator, i.e., gender. This argument is countered by a significant conditional effect of commute on total CFP at various values of the moderator. Thus, gender has significant conditional impact on predicting total CFP (p-values < 05).

The total commute time is a significant predictor of total CFP (p-value < 0.001). The moderator, gender, is non-significant predictor of total CFP ($p = 0.79$). The interaction effect is non-significant ($p = 0.49$). This means as total commute time increases, the increased value of total CFP may not be due to the moderator, i.e., gender. This argument is countered by a significant conditional effect of total commute time on total CFP at values of the moderator, i.e., gender (p-values < 0.05). Thus, the variable gender has significant conditional impact on predicting TOTAL CFP.

The kilometer distance from home and designation are significant predictors of total CFP (p-values < 0.05). The Interaction effect is significant ($p = 0.015 < 0.05$), means as kilometer distance from home is more for the commuters the increased value of total CFP may be due to the moderator, i.e., designation. This argument is further supported by conditional effect of kilometer distance from home on total CFP at values of the moderator, i.e., designation (p-values < 0.05). The kilometer distance from home is a significant predictor of total CFP conditional on the various values of moderator variable, designation. Thus, designation has a significant conditional impact on predicting total CFP.

The time of commute is a significant predictor of total CFP ($p = 0.0001$). The moderator, designation is a non-significant predictor of total CFP ($p = 0.062$). The interaction effect between them is non-significant ($p = 0.07$). This means as commute time increases, the increased value of total CFP may not be due to the moderator, i.e., designation. This argument is further countered by significant conditional effect commute time on total CFP at various categories of the moderator, i.e., designation (p-values < 0.05). Thus designation has a significant conditional impact on predicting total CFP.

The kilometer distance from home and campus location are significant predictors of total CFP (p-values < 0.05). The Interaction effect between them is significant ($p = 0.016$). This means as kilometer distance from home increases, the increased value of total CFP may be due to the moderating variable, i.e., campus location. This argument is further supported by the conditional effect of kilometer distance from home on total CFP at various categories of the moderator, i.e., campus location (p-values < 0.05). Thus, the moderator, campus location has a significant conditional impact on predicting total CFP.

The commute time, campus location and the interaction effect between them are significant predictors of total CFP (p-values < 0.05). The interaction effect is significant ($p < 0.001$), means as the commute time increases, the increased value of total CFP may be due to the moderator, i.e., campus location. This argument is further supported by the conditional effect of commute time on total CFP at various values of the moderator, i.e., campus location (p-values < 0.05).

3.6 Discussion on Employee Commuters CFP Reduction

Based on the above analysis, descriptive statistics, qualitative inputs of the survey, and discussions conducted with employees, the following groups of employees are formed, aimed at suitable future CFP reductions. These groupings are shown in Table 3.3. A useful sample of 350 was selected for this grouping, excluding 5 employees who had fixed choices on train commutes.

3.6.1 Group 1: 37.2% of Symbiosis Employees Commute Exclusively on Two-Wheelers

Based on the demography of employees, more administrative and support staff use two-wheelers than executives and faculty. The common reasons for the use are affordability in capital investment for vehicle purchase, cheaper

Table 3.3   Grouping employee commuters

Group	Percentage (of Sample Size = 350)	Count of Employees (Sample Size = 350)	Method of Commuting	Their Relevant Share of Carbon Footprint (CFP) Out of TOTAL Symbiosis Carbon Footprint (Metric Tons)	Employee CFP per Capita (Metric Tons)
1	37.80%	132	exclusively on two wheelers	24.94% of TOTAL Symbiosis CFP (66.94 metric tons)	0.51 (66.94 metric tons/ 132 employees)
2	24.30%	85	exclusively by bus	25.77% of TOTAL Symbiosis CFP (69.17 metric tons)	0.81
3	15.50%	54	exclusively by car	20.74% of TOTAL Symbiosis CFP (55.66 metric tons)	1.03
4	12.00%	42	Short distance by two-wheeler, then by bus	17.75% of TOTAL Symbiosis CFP (47.64 metric tons)	1.13
Total	89.2% of employees	313 out of 350 sample		89.20% of TOTAL Symbiosis commuter CFP (268.40 metric tons for 350 sample)	0.756 is average CFP per capita for 350 sample

Source: Authors.

fuel costs, and vehicle operational costs. Two wheelers were observed to be the most common method of travel throughout Pune city, similar to most urban India locations (Krishnamurthy et al., 2016; Kamble et al., 2009; Maunder et al., 1997). Currently at SIU, the sharing of two wheelers is informal and depends on the vehicle owner's friendship circles, based on qualitative responses of survey and selected employee interviews. Additionally, 29% of all two-wheeler commuters practice vehicle-sharing, mostly male commuters, indicating well-utilized CFP reduction potential. In the conservative culture of India, same gender sharing of two-wheelers is socially acceptable. If this prevalent mode of employees' transport serves as example-setting, a further 21% of two-wheeler commuters who responded willingness to share could be encouraged to participate in two-wheeler sharing. Apart from adding more employees, this encouragement could help further reduce employee CFP from an already low per capita CFP of 0.51 metric tons. This view is further supported by regression that irrespective of the distance from home or commute time increases; the increased CFP may not be due to the choice of two wheeler mode. If implemented systematically, further technological assistance by mobile applications and data sharing could enable employees to connect for vehicle sharing. Two-wheeler sharing could reduce many inconveniences of commuters' time on the road and could encourage social bonding.

3.6.2 Group 2: 24.3% of Symbiosis Employees Commute Exclusively by Bus

A majority of these employees live close to bus pick up point (average 1 km) or prefer to walk to a campus bus pick up point. An interesting note in this category is that 34% of them mentioned they would not consider changing to Pune city buses. Qualitative responses of the survey indicate that employees experienced convenience in the drop off and pick up within campus at scheduled 9 am to 5 pm campus work hours, thus avoiding peak hour traffic congestions and door-step service at one end. Women employees emphasized their priority for domestic care after 5 pm and appreciated safety concerns alleviation (Levy, 2013). Standard work timings are practiced less in other business organizations, where overtime work is common. For these stated reasons, limited commuting options were acceptable for employees. Several employees of Viman Nagar and Hinjewadi campuses mentioned traffic congestion-related delays and discomforts. Very few employees of all seven campuses mentioned on-road inconveniences, even though all the campus buses are non-air-conditioned. Many employees in the lower and higher cadres alike preferred buses, since it provided opportunities to unwind after work and chat with co-workers during the commute. Although the average CFP of a diesel powered bus is higher, this group is more carbon-friendly due to CFP sharing among 30–50 people per bus. Their CFP is certainly lower than exclusively-by-car commuters, despite shorter distances commuted by car.

3.6.3 Group 3: 15.5% of Symbiosis Employees Commute Exclusively by Car

The demographic profile of the car commuters indicates a majority of them are teaching faculty and officers, pertinent to higher socio-economic status. Symbiosis employees need to be encouraged to share their cars more. The same was an observed trend in many urban locations in Pune city, as per an earlier observation study by SIU students. As confirmed by regression, for car commuters, as the distance or commute time increases, the CFP increases significantly. Thus, solo car commuters contribute the maximum CFP, in comparison to all categories. Qualitative responses of the survey indicate that although affordability exists among such employees, the key reason to avoid car-sharing seems to be inconvenience in rerouting for other pickups, maintaining punctuality in carpooling habits and privacy preferences. The secondary benefits expected were reduced stress while driving, and formation of social bonds. Other organizations reporting carpooling indicate usage of dedicated company websites and developing informal groups among employ-ees (Paladugula and Rathi, 2013). These observations lead to future scope of research i.e. the investigation into barriers for car sharing. It may be argued that until high-consequences risk is felt (for instance, petrol supply shortages or very high petrol prices) commuters may not feel encouraged to share. An overwhelming majority of vehicles (99%) owned petrol engine cars, the rest being diesel or natural gas. 28% of overall employees were willing to consider taxi-pooling as an option, indicating good potential for CFP reduction. As per qualitative inputs of Survey, location-based hi-tech bus services can be encouraged. Such services are available for IT company zones in Pune, but a subscription for the same would cost more than campus bus transit facility. It may be argued that national fuel price subsidies for all sectors would be well utilized if car-pooling is widely adopted in Urban India commutes.

3.6.4 Group 4: 12.0% of Symbiosis Commuters Travel by Two-Wheeler and then by Bus

A majority of these employees make a short travel to a campus bus pick up point (average 5.7 km commuted in 16 minutes) and then complete remaining travel by campus bus (average 16.3 km commuted in 39 minutes). The comparatively higher CFP is attributed to longer bus-commute distances. CFPs of diesel engine buses are larger, but when shared among 30 occupants, it reduces the CFP per capita employee. The campus bus transit services already encourage much sharing activity to promote efficiency and convenience. Currently, this unintended CFP reduction co-benefits could be purposely repositioned to reduce diesel fuels usage and phase out to natural gas or hybrid bus engines (Sharma and Tomar, 2010). To enhance carbon-friendly practices, previously mentioned sharing and organizing newer options need to be experimented. LCE practices require trial and error approaches, which allow time and generate acceptance by commuters to facilitate desirable emergent practices.

3.7 Recommendations

3.7.1 ICT-Enabled Commuter to Commuter Connections for Vehicle Sharing

Technology dependencies such as mobile applications are becoming the norm in all services. SIU could promote CFP reduction by adopting mobile applications in the following ways: (i) to check available bus routes and congestions. (ii) Bike taxi companies have recently begun connecting willing partners to share bike rides in the city. These can be customized to partner willing campus commuters. (iii) Taxi-pooling is a popular cost savings option in Indian cities. Cab companies manage logistics by mobile applications, thus drastically enabling CFP reductions of urban commuters. A Dedicated vehicle-sharing website can be designed, with employees' details of: preferred locations for pick up, drop off, usual work timings, availability of vehicles to share, and wish to share details. Previously unknown possibilities of vehicle sharing can be fostered, when employees willingly share their logistics to identify SIU employees for vehicle sharing in their neighborhood. Frequent vehicle sharing can be encouraged. However, such commitments to carpooling habits require tolerance for minor inconveniences in rerouting and privacy.

3.7.2 Private Mass Rapid Transport

Some traffic congested regions have developed their regional private partnerships to provide comfortable and convenient commutes for their employees. These services require higher monthly membership fees, the commuter's employment within the relevant city region, and demands the employer to be a registered industry association member. Such options throw open a wide variety of timings and route choices closer to door-step than individual organizations' transit facilities. Such private company fleets are observed to be successful with satisfied high membership and profitable operations. When coordinated well, conversion of such fleets from diesel fueled to hybrid vehicles is a promising future possibility for CFP reductions capability among public private partnerships (Devkar et al., 2013).

3.8 Conclusion

The potential of HEIs towards LCE was explored with a chosen University at Pune, an educational hub in India. Swift urbanization of many Indian cities is a motivation to study this sector's potential for CFPs reduction. This chapter demonstrates data collection and CFP estimation possibilities for other Indian HEIs. The grouping of vehicles and commute patterns may vary among Indian cities. However, the intention to develop awareness within HEIs is of immense importance in India today, since several possibilities for further research and remedial measures arise from baseline studies. This commuters CFP assessment aspires to make a case for replicable baseline studies within many HEIs and organizations in India. Global connections with other cities are a crucial need of the hour since many Indian cities such as Pune suffer from imbalanced growth in traffic and congestions. Adequate transport infrastructure development requires extensive participation from urban HEIs of India. Anticipating future technological dependences, a few proactive measures to ease CFP in an employee friendly manner were recommended. Researched disseminations of appropriate mitigation measures by HEIs could very well motivate other HEIs. Such widespread disseminations will inspire pan-India sustainability practices adoptions to reduce CFPs within organizations.

Acknowledgement

This research project was funded by Symbiosis International University minor research grant. The authors gratefully acknowledge SIU, the support extended by several departments during data collection, and the various Symbiosis employees for their multiple inputs in developing this study.

References

Aleixo, A. M., Azeiteiro, U. M., and Leal, S. (2017). "UN decade of education for sustainable development: perceptions of higher education institution's stakeholders," in *Handbook of Theory and Practice of Sustainable Development in Higher Education*, eds Leal Filho W., Azeiteiro U., Alves F., and Molthan-Hill P (Basel: Springer International Publishing), 417–428.

Aroonsrimorakot, S., Yuwaree, C., Arunlertaree, C., Hutajareorn, R., and Buadit, T. (2013). *Carbon Footprint of Faculty of Environment and Resource Studies*, (Nakhon Pathom: Mahidol University).

Beermann, J. (2017). *Urban Cooperation and Climate Governance: How German and Indian Cities Join Forces to Tackle Climate Change*. Berlin: Springer.

Beermann, J., Damodaran, A., Jörgensen, K., and Schreurs, M. A. (2016). Climate action in Indian cities: an emerging new research area. *J. Integr. Environ. Sci.* 13, 55–66.

Clifford, J. M., and Cooper, C. D. (2012). A 2009 mobile source carbon dioxide emissions inventory for the University of Central Florida. *J. Air Waste Manag. Assoc.* 62, 1050–1060.

Dash, N., and Balachandra, P. (2016). Benchmarking Urban sustainable efficiency: a case of indian cities. *Trans. Res. Proc.* 14, 1809–1818.

Dasmann, R. F. (1988). "Toward a biosphere consciousness," in *The Ends of the Earth: Perspectives on Modern Environmental History*, eds D. Worster, (Cambridge: Cambridge University Press), 277–288.

Devkar, G. A., Mahalingam, A., and Kalidindi, S. N. (2013). Competencies and urban public private partnership projects in India: a case study analysis. *Policy Soc.* 32, 125–142.

Disterheft, A., Caeiro, S., Azeiteiro, U. M., and Leal Filho, W. (2015). Sustainable universities – a study of critical success factors for participatory approaches. *J. Clean. Produc.* 106, 11–21.

Doll, C. N. H., Dreyfus, M., Ahmad, S., and Balaban, O. (2013). Institutional framework for Urban development with co-benefits: the Indian experience. *J. Clean. Produc.* 58, 121–129.

Duffield, C., Gardner, J. S., Berkes, F., and Singh, R. B. (1998). Local knowledge in the assessment of resource sustainability: case studies in

Himachal Pradesh, India, and British Columbia, Canada. Mt. Res. dev. 18, 35–49.

Dutt, A. K., Noble, A. G., Costa, F. J., Thakur, R. R., Thakur, S. K., and Sharma, H. S. (2016). *Spatial Diversity and Dynamics in Resources and Urban Development*. Dordrecht: Springer.

Fatima, E., and Kumar, R. (2014). Introduction of public bus transit in Indian cities. *Int. J. Sustain. Built Environ*. 3, 27–34.

Finlay, J., & Massey, J. (2012). Eco-campus: applying the ecocity model to develop green university and college campuses. *Int. J. Sustain. High. Educ*. 13, 150–165.

Garg, A., Bhattacharya, S., Shukla, P. R., and Dadhwal, V. K. (2001). Regional and sectoral assessment of greenhouse gas emissions in India. *Atmos. Environ*. 35, 2679–2695.

Gupta, D. K., Bhatia, A., Kumar, A., Das, T. K., Jain, N., Tomer, R., et al. (2016). Mitigation of greenhouse gas emission from rice–wheat system of the Indo-Gangetic plains: Through tillage, irrigation and fertilizer management. *Agric. Ecosyst. Environ*. 230, 1–9.

Hansson, C. B., and Wackernagel, M. (1999). Rediscovering place and accounting space: how to re-embed the human economy. *Ecol. Econ*. 29, 203–213.

India GHG Program (2015). *Carbon Footprint on-Line Calculation Tools*. Available at: http://indiaghgp.org/

Jha, R. (2015). *A Study of Urban Planning and its Implications for Improving Living Conditions in Maharashtra State*. Ph.D. thesis, Tilak Maharashtra Vidyapeeth, Pune.

Jörgensen, K., Mishra, A., and Sarangi, G. K. (2015). Multi-level climate governance in India: the role of the states in climate action planning and renewable energies. *J. Integr. Environ. Sci*. 12, 267–283.

Kamble, S. H., Mathew, T. V, and Sharma, G. K. (2009). Development of real-world driving cycle: Case study of Pune, India. *Trans. Res. D* 14, 132–140.

Kantakumar, L. N., Kumar, S., & Schneider, K. (2016). Spatiotemporal urban expansion in Pune metropolis, India using remote sensing. *Habitat Int*. 51, 11–22.

Kathuria, A., Parida, M., Ravi Sekhar, C., and Sharma, A. (2016). A review of bus rapid transit implementation in India. *Cogent Engin*. 3:1241168.

Krishnamurthy, R., Mishra, R., & Desouza, K. C. (2016). City profile: Pune, India. *Cities* 53, 98–109.

Krishnamurthy, S., Joseph, S., and Bharathi, V. (2014). Creating environment friendly projects in rural India – A synergy framework for sustainable renewable energy. *Int. J. Appl. Engin. Res*. 9, 26719–26738.

Kundu, D. (2014). Urban development programmes in India: a critique of JnNURM. *Soc. Change* 44, 615–632.

Lewis, J. I., and Wiser, R. H. (2007). Fostering a renewable energy technology industry: an international comparison of wind industry policy support mechanisms. *Energy policy* 35, 1844–1857.

Levy, C. (2013). Travel choice reframed: "deep distribution" and gender in urban transport. *Environ. Urban.* 25, 47–63.

Mane, T. T., and Hingane, H. N. (2012). Existing situation of solid waste management in Pune City, India. *Res. J. Rec. Sci.* 1, 348–351.

Maunder, D., Palmner, C., Astrop, A., Babu, M., Maunder, D., Palmer, C., et al. (1997). Attitudes and travel behaviour of residents in Pune, India. *Paper presented at the Annual Meeting of the Transportation Research Board,* Washington, DC.

Messner, D., Schellnhuber, J., Rahmstorf, S., & Klingenfeld, D. (2010). The budget approach: A framework for a global transformation toward a low-carbon economy. *J. Renew. Sustain. Energy* 2:031003.

MOUD (2017). *Ministry of Urban Development, Government of India (GOI)*, Available at: http://moud.gov.in/cms/JNNURM.php and https://india.gov.in/official-website-jawaharlal-nehru-national-urban-renewal-mission [accessed at May 10, 2017].

Mulugetta, Y., and Urban, F. (2010). Deliberating on low carbon development. *Energy Policy* 38, 7546–7549.

Nuzir, F. A., and Dewancker, B. J. (2014). Understanding the role of education facilities in sustainable urban development: a case study of KSRP, Kitakyushu, Japan. *Proc. Environ. Sci.* 20, 632–641.

Paladugula, A. L., and Rathi, S. (2013). Strategies to reduce energy use for commuting by employees. *Proc. Soc. Behav. Sci.* 104, 952–961.

Panwar, N. L., Kaushik, S. C., & Kothari, S. (2011). Role of renewable energy sources in environmental protection: a review. *Renew. Sustain. Energy Rev.* 15, 1513–1524.

Pradoto, W. (2012). *Development Patterns and Socioeconomic Transformation in Peri-Urban Area.* Berlin: Technische Universitet Berlin.

Pucher, J., Korattyswaropam, N., Mittal, N., and Ittyerah, N. (2005). Urban transport crisis in India. *Transp. Policy.* 12, 185–198.

Ramachandra, T. V., and Shruthi, B. V. (2007). Spatial mapping of renewable energy potential. *Renew. Sustain. Energy Rev.* 11, 1460–1480.

Rao, P. (2011). "Information and communication technologies (ICT) in building knowledge processes in vulnerable ecosystems: a case for sustainability," in *Green Finance and Sustainability: Environmentally-Aware Business Models and Technologies*, (Hershey, PA: IGI Global), 176–190.

Rao, P., and Leal Filho, W. (2015). "Local community perception of climate change and scientific validation: a review of initiatives and perspectives in the indian region," in *Climate Change in the Asia-Pacific Region,* ed. W. Leal Filho (Cham: Springer International Publishing), 89–101.

Revi, A. (2008). Climate change risk: an adaptation and mitigation agenda for Indian cities. *Environ. Urban.* 20, 207–230.

Rode, P., Floater, G., Thomopoulos, N., Docherty, J., Schwinger, P., Mahendra, A., et al. (2014). "Accessibility in cities: transport and Urban," in *Disrupting Mobility. Lecture Notes in Mobility,* eds Meyer G., and Shaheen S. (Cham: Springer), 1–61.

Roy, R., Potter, S., and Yarrow, K. (2008). Designing low carbon higher education systems: environmental impacts of campus and distance learning systems. *Int. J. Sustain. High. Educ.* 9, 116–130.

Sapkota, T. B., Jat, M. L., Aryal, J. P., Jat, R. K., and Khatri-Chhetri, A. (2015). Climate change adaptation, greenhouse gas mitigation and economic profitability of conservation agriculture: Some examples from cereal systems of Indo-Gangetic Plains. *J. Integr. Agric.* 14, 1524–1533.

Saroha, J. (2016). "Sustainable urbanization in India: experiences and challenges," in *Spatial Diversity and Dynamics in Resources and Urban Development* (Dordrecht: Springer), 81–98.

Saveyn, B., Paroussos, L., and Ciscar, J. C. (2012). Economic analysis of a low carbon path to 2050: a case for China, India and Japan. *Energy Econ.* 34, S451–S458.

Schmid, G. (2012). The development of renewable energy power in India: which policies have been effective? *Energy Policy* 45, 317–326.

Sellers, J., Han, S. S., Huang, J., Lu, X. X., Marcotullio, P., and Ramachandra, T. V. (2009). *Peri-Urban Development and Environmental Sustainability: Examples from China and India.* Kobe: Asia-Pacific Network for Global Change Research.

Sharma, D., & Tomar, S. (2010). Mainstreaming climate change adaptation in Indian cities. *Environ. Urban.* 22, 451–465.

Shen, L. Y., Ochoa, J. J., Shah, M. N., and Zhang, X. (2011). The application of urban sustainability indicators–A comparison between various practices. *Habitat Int.* 35, 17–29.

Shirgaokar, M. (2014). Employment centers and travel behavior: exploring the work commute of Mumbai's rapidly motorizing middle class. *J. Transp. Geogr.* 41, 249–258.

Shukla, P. R., and Chaturvedi, V. (2012). Low carbon and clean energy scenarios for India: analysis of targets approach. *Energy Econ.* 34, S487–S495.

Smith, K. R., Uma, R., Kishore, V. V. N., Zhang, J., Joshi, V., and Khalil, M. A. K. (2000). Greenhouse implications of household stoves: an analysis for India. *Annu. Rev. Energy Environ.* 25, 741–763.

Spirovski, D., Abazi, A., Iljazi, I., Ismaili, M., Cassulo, G., and Venturin, A. (2012). Realization of a Low Emission University Campus Trough the Implementation of a Climate Action Plan. *Proc. Soc. Behav. Sci.* 46, 4695–4702.

Subbarao, S. S. V, & Krishna Rao, K. V. (2013). Trip chaining behavior in developing countries: a study of Mumbai Metropolitan Region, India. *Eur. Transp.*

Venkataraman, C., Habib, G., Eiguren-Fernandez, A., Miguel, A. H., and Friedlander, S. K. (2005). Residential biofuels in South Asia: carbonaceous aerosol emissions and climate impacts. *Science* 307, 1454–1456.

West, J. J., Smith, S., Silva, R. A., Adelman, Z., Fry, M. M., Anenberg, S., et al. (2011). "Co-benefits of global greenhouse gas mitigation for air quality and human health via two mechanisms," in *Proceedings of the AGU Fall Meeting*, San Francisco, CA.

Yaduvanshi, N. R., Myana, R., and Krishnamurthy, S. (2016). Circular economy for sustainable development in India. *Indian J. Sci. Technol.* 9, 1–9.

4

Progressive Trends in Implementing Climate Change Courses in Higher Education Curriculum at Symbiosis International University, Pune, India

Prakash Rao[1] and Yogesh Patil[2]

[1]Department of Energy and Environment,
Symbiosis Institute of International Business,
Symbiosis International University, Pune, Maharashtra, India
[2]Symbiosis Centre for Research and Innovation,
Symbiosis International University, Pune, Maharashtra, India

Abstract

Purposes – In recent times the issue of climate change is seen as perhaps the most important global environmental threat affecting the very fabric of society and economic growth. This is likely to have significant impacts on ecosystem, natural resources, local livelihoods and other sectors. From a business perspective serious challenges are seen as many industries rely on natural resources for production and operations.

Methodologies – The Symbiosis International University through its Sustainability policy intends to increase and improve the environmental content of curricula at its constituent institutes by developing curriculum for undergraduate students and post graduate students focusing on sustainable technologies, management, and policy and advocacy issues.

Results – The curriculum using the centrality approach also delves into UN policy processes of mitigation and adaptation following the post Paris (COP21) climate negotiations. The study evaluates current trends in course curriculum development in line with Global climate negotiations as well as national climate priorities.

Implications – The pedagogical approach uses assessments of sectoral Greenhouse Gas (GHG) emissions through class projects and real time exercises using the Institute emissions as a live example. The courses are also tied to innovative market based mechanisms and bring in recent industry technologies and regulatory driven approaches like the PAT, REC schemes, CERC etc.

Value – Low carbon development based curriculum within courses in the management discipline through the direct linkages between global climate change impacts and energy development is seen as an enabler for students to demonstrate low carbon activities through class room courses as well as in practice.

4.1 Introduction

Climate Change is now recognised as a potential threat to sustainable livelihoods and our natural ecosystems (Battisti and Naylor, 2009). Since the last few decades the average surface temperature of the earth have shown a steady increase with Inter Governmental Panel on Climate Change (IPCC) indicating that average global temperatures likely to rise by 1.1–6.4°C over the next few years (IPCC, 2014). The consequences of such warming have resulted in atmospheric changes leading to extreme weather events like cyclones, storms, droughts, etc. This phenomenon can have serious implications on biological diversity (Lovejoy and Hannah, 2006) as well as significant social and economic impacts on the livelihood patterns of people particularly those living in the tropical regions (Agarwal and Narain, 1991). Some of the probable impacts that have been identified include changes in temperature and rainfall regime, agriculture yields, sea level rise (Byravan et al., 2012), extreme weather events, melting of glaciers (WWF, 2005) changes in ecosystem function (Jindal and Mankotia, 2004) forest fires etc.

Changes in habitat ranges and shifts in vegetation types could affect the distribution patterns of these important species as a result of local climate variability. This is also likely to affect the Protected Areas system in the state as a result of faunal range movement. Further studies are necessary to document the ecological changes taking place in the region and consequent shifts in ranges and habitats of key mammalian species.

Objectives

- To review the development of academic courses in the context of evolving climate change dynamics at a progressive University.
- To assess the centrality approach for curriculum development using climate linked courses and topics.

4.2 The Global Situation

During the 1980s, scientific evidence about the possibility of global climate change led to growing public concern. By 1990, a series of international conferences had issued urgent calls for a global treaty to address the problem. The United Nations Environment Programme (UNEP) and the World Meteorological Organization (WMO) responded by establishing an intergovernmental working group to prepare for the treaty negotiations.

In response to the working group's proposal, the United Nations General Assembly at its 1990 session set up the Intergovernmental Negotiation Committee for a Framework Convention on Climate Change (INC/FCCC) (Dasgupta, 2012). After a series of meetings and negotiations a high profile Kyoto (www.unfccc.int) conference resulted in a consensus decision to adopt a protocol under which industrialized countries would reduce their combined greenhouse gas emissions by at least 5% compared to 1990 levels by 2012. This legally binding commitment promised to produce a historic reversal of the upward trend in emissions that started in these countries some 150 years ago (Muller et al., 2009).

4.2.1 India's Participation in International Negotiations and Climate Policy

India as one of the major developing nations has taken a very proactive stance in the process of international negotiations of climate change. India earlier ratified the Kyoto Protocol and more recently the Paris agreement of 2015 (Parikh, 1995; Jakobsen, 1998).

Being a major developing nation India has mobilized support to bring adaptation as an important issue in addressing climate change (Sengupta, 2012). The country used this to maximum advantage when it hosted the Eighth Conference of Parties (COP 8) of the UNFCCC at New Delhi during October 2002. Though the country has opposed any legally binding commitment to emission reduction, it has taken various policy level measures and on ground activities which will lead to the reduction of greenhouse gas emissions (Rajamani, 2007; IEA, 2007).

4.2.1.1 International climate policy

In today's changed scenario, India has stand high in the comity of nations where which have successfully campaigned for low carbon world through effective negotiations at various UNFCCC meetings and conferences. Through meaningful initiatives and a few international and regional activities, the country aims to address the climate change issue in global arena and corresponding activities locally, to support the global efforts, to reduce greenhouse gas emission without disturbing the pace of development and growth. The aim is to influence and drive the international negotiation process through the mechanism of INDCs and through nationally approved climate action plans. This will help in achieving the overall goal of managing and limiting the impacts of global warming in a most effective and rational way.

4.2.2 Promote Sustainable Energy Practices (SEP)

Energy consumption is considered one of the key development indicators of a country. In India, though the per-capita energy consumption is much lower than the world average, the overall energy consumption is substantial and increasing at an exponential rate. The country's power policy and planning lacks transparency, and mostly overlooks environmental and social implications, leading to unsustainable growth. Presently, India is on a steep growth curve (average GDP is increasing at a rate ~7%), and ranks sixth in the world in terms of energy requirement. Though the per capita energy consumption in the country is far below the world average, the overall energy consumption is significant. The power sector meets a major share of energy requirements of the country's expanding economic growth. The power generation in the country is coal intensive and is resulting in significant amounts of CO_2 emissions, threatening the socio-economic and ecological health of the country (MoEF, 2010) There is an urgent need to effectively address the country's development priorities and energy requirements as a whole (Sant and Gambhir, 2012) to achieve a sustainable development and future (Sreekumar and Dixit, 2010).

4.3 Climate Change in Higher Education Curriculum

Climate change is an emerging issue and in developing countries like India, where the country is more concerned about achieving food security, energy security there is very little awareness about the subject in primary and secondary educational institutes including higher education institutions (HEI). There is a lack of capacity coupled with low levels of awareness about climate change and its impacts on biodiversity and livelihoods (Leal Filho, 2011, 2012). At the 7th Conference of the Parties to the United Nations Framework Convention on Climate Change (UNFCCC) in Marrakesh, parties identified and strengthened the framework, and the need for capacity building of the developing countries on climate change issues.

Building capacities of Community Based Organization (CBOs) and Civil Society should act as catalyst at the local level for climate change vulnerability assessment and subsequent identification of issues for integrating it into development planning and policy. Capacity building of local government, local self-governing bodies, industries, HEI, etc. will facilitate a gradual seeding of the intervention and empower communities and promote an integrated development model.

Industries and corporates now see investing in environment protection efforts not only as a business proposition but also in terms of significant economic gains by way of cost reduction and optimisation of resource use and reduced carbon footprint. The recent climate change negotiations have brought to the fore some of the important industry initiatives in promoting clean technologies, processes and capacity building efforts. Various Industry associations in India like CII, FICCI, ASSOCHAM have strongly advocated environmental responsibility as a key mandate of business operations and sustainable practices (Pulver, 2002).

The industry today is focusing on emerging areas like renewable energy, energy efficiency, R&D for innovative and clean technologies, sustainable energy options, sustainable transport and infrastructure development, green building architecture etc. which has taken centre stage in the development of better environment practices.

4.3.1 Internationalisation of Global Climate Change as an Integrated Curriculum at SIIB

In 2009, the Symbiosis International University, Pune through its constituent institute the Symbiosis Institute of International Business, developed an innovative post graduate programme which was aimed at integrating environmental sustainability, climate change, energy production, and economic growth in its curriculum (Rao et al., 2013). The programme provided for a holistic understanding of developing competencies in emerging technologies, economic issues and global environmental strategies. Core focus areas include sectors like sustainable energy development, renewable energy, power economics, climate change, carbon financing and markets, corporate sustainability, environmental impact assessments, natural resources management, etc.

The programme was conceived in the backdrop of growing interest in globalisation as part of the Government of India National Action Plan on Climate Change (NAPCC) on developing an understanding of global climate change and its impacts on India's ecosystems, as well as to come up with mitigation efforts and to initiate adaptive responses in the face of adverse climate change (IPCC, 2007). As a post-graduate business school it was also essential to develop a curriculum which encapsulated the essence of climate change through specific courses which were focussed on the link between climate change, energy and business sustainability. It may be mentioned here that SIIB was perhaps one of the first Post graduate institutes in the country which developed an integrated curriculum centred around climate change (Rao and Patil, 2015) by including several courses which had relevance to industry energy linkage and mitigation efforts. Creation of industry based courses on climate change was also driven to a large extent by the UNFCCC where market mechanisms like the Clean Development Mechanism (CDM) saw the emergence of business opportunities for project developers, investors, assurance agencies, advisory services groups, trading and brokerage groups, policy makers, research scientists, civil society and social scientists. In the context of the first phase of the Kyoto Protocol this was seen as a game changer as it enabled developing countries (non Annex 1 Countries) to establish clear priorities towards sustainable development which was also one of the goals of the CDM process (UNFCCC, 1992). The recent UN Conference on Sustainable development (UNCSD, 2012) has further stressed the need for evolving future strategies around how educational institutions and universities can provide direction to sustainable development mechanisms through various initiatives like sustainability curriculum development, green technologies, environment awareness campaigns etc. In such a scenario, SIIB saw the introduction of climate oriented curriculum, the need of the hour (Table 4.1). As an institute with a focus on International business, it was felt important to include industry relevant courses focusing purely on climate change and energy given its international context.

Table 4.1 Climate change centered curriculum at Symbiosis Institute of International Business (SIIB)

S. No.	Sector
1.	Climate change and energy development
2.	Energy conservation
3.	Power economics, governance and energy markets
4.	Renewable energy sources

4.4 The Evolving Climate Change Curriculum at SIIB

The concept of designing low-carbon universities not only through sustainable development-based curriculum and frameworks (Chambers, 2009; Viebahn, 2002) but also through sustainable operations and research on innovative technologies has evolved considerably in recent time. These are evident through studies on the environmental impacts of campuses and educational institutions. Several universities have also initiated research and action to assess their GHG inventory (University of Toronto, 2010) and tracking their carbon emissions profile (Sprangers, 2011).

Since the introduction of environment as a subject across post graduate education institutions in India in 1985, sustainable development issues at Universities and academic institutions in India has seen sufficient academic interest from the point of view of Institutions introducing an elective based courses on climate change, carbon foot printing, climate policy or comprehensive programmes at UG or PG levels (Rao, 2011; Rao and Patil, 2015). The role of business schools in creating leadership to tackle sustainability challenges has been a key factor (Adams et al., 2011; Amran et al., 2010) in driving the development of sustainable or low carbon universities.

Several authors have emphasized on the need to look at sustainability at campuses from an education and capacity building (Adams, 2011; Amran et al., 2010; Coops et al., 2015; Godemann et al., 2011; Naeem and Neale, 2012; Rao, 2011; Rao and Patil, 2015; Shriberg, 2012), research (Sprangers, 2011), campaigns and awareness (Stubbs, 2013; Stubbs and Cocklin, 2008), resource use (Pandey et al., 2011; Patil and Rao, 2014; Roy et al., 2008) and policy (Carbon Trust, 2006) perspective.

When the integrated Energy and Environment programme at SIIB was conceived in 2009 with the support of industry representatives, it was strongly felt that an integrated curriculum covering all aspects of energy and environment should be the core focus of this unique programme. In the initial years, this was limited to conducting courses on carbon credits and energy conservation.

Climate change was recognised as an overarching global problem and perhaps the most defining environmental issue of the 21st century. With the formation of the UNFCCC in 1992 and the scientific and unequivocal evidence linking human interference to climate change impacts (IPCC, 2014), there was need to instil international business education curriculum related to this environmental issue and through critical class room discourses of Internationalisation of the subject and global policy considerations.

The University in 2013 embarked on a curriculum development exercise aimed at standardisation its curricula across various faculty disciplines including the sustainability and infrastructure area. During this effort a Sub-committee comprising of faculty members drawn from sustainability area, as also external experts was formed to review the evolving course keeping in mind the rapid international developments in the climate change sector. The effort was meant to integrate several courses which otherwise were seen as stand-alone courses with no proper linkages across climate change and energy.

The programme during the reorganisation process developed niche based courses with specific focus with a view to integrate some of the critical issues linking global climate change with sustainable energy development as well as linkages with adaptation and mitigation of climate impacts (Table 4.2).

The reorganised course curriculum was in line with International and global climate dynamics and necessitated that two important areas should be considered for business education curriculum at SIIB.

4.4.1 Climate Change and Energy Development

The SIIB programme tried to integrate a low carbon development based curriculum (Date-Huxtable et al., 2013) within courses in the management discipline through the direct linkages between global climate change impacts and energy development but also as an enabler for students to demonstrate low carbon activities through practice. The pedagogical approach used assessments of sectoral Greenhouse Gas (GHG) emissions through class projects and real time exercises using the Institute emissions as a live example. This enabled students to integrate local-national-global climate related issues with GHG-Energy use-low carbon economy. The curriculum using the centrality approach also analysed UN policy processes of mitigation and adaptation following the post Paris (COP21) climate negotiations including an understanding of the Intended Nationally Determined Contributions (INDCs) for signatories to the UNFCCC.

Table 4.2 Climate and Energy Courses at SIIB

S. No.	Courses	2009–11	2010–12	2011–13	2012–14	2013–15	2014-16	2015–17	2016–18
A	Carbon Financing and Project Design Document	–	–	√	√	–	–	–	–
A	Carbon Financing and Market Instruments	–	–	–	–	√	–	–	–
A	Climate Change	√	√	–	–	–	–	–	–
A	Climate Change Impacts, Mitigation and Adaptation	–	–	√	–	√	–	–	–
A	Climate Change, Carbon Markets and Financing	–	–	–	–	–	√	√	√
A	Understanding Carbon credits	–	–	–	√	√	–	–	–
A	Understanding Carbon Credits and Markets	–	√	√	–	–	–	–	–
B	Energy Asset Management	√	–	–	–	–	–	–	–
B	Energy Audits	√	–	–	–	–	–	–	–
B	Energy Audits and Energy Conservation	–	√	–	–	–	–	–	–
B	Energy Audits and Energy Management	–	–	√	√	√	√	√	√
B	Energy Conservation	√	–	–	–	–	–	–	–
B	Energy Conservation and Sustainable Energy Development	√	–	–	–	–	–	–	–
B	Introduction to Energy Regulation and Policies	–	√	√	√	√	√	√	√
B	Non-Renewable Energy Sources	√	√	√	√	√	√	√	√
C	Contracting in Energy Sector	–	–	–	–	–	√	√	√

(Continued)

Table 4.2   Continued

S. No.	Courses	2009–11	2010–12	2011–13	2012–14	2013–15	2014–16	2015–17	2016–18
C	Energy Derivatives and Risk Management	–	–	✓	–	–	–	–	–
C	Energy Economics	✓	✓	–	–	–	–	–	–
C	Energy Marketing/Power Trading	✓	✓	–	–	–	–	–	–
C	Energy Modelling	✓	✓	–	–	–	–	–	–
C	Energy Project Appraisal and Financing	–	–	–	✓	✓	–	–	–
C	Financing the Energy Sector	✓	–	–	–	–	–	–	–
C	Power Trading	–	–	✓	–	–	–	–	–
C	Power Trading and Energy Markets	–	–	–	–	✓	✓	–	–
C	Project Financing in Energy Sector	–	✓	–	–	–	–	–	–
D	Electricity Regulatory Issues and Mechanisms	–	–	–	–	–	✓	✓	✓
D	Power Economics and Regulation	–	–	✓	–	–	–	–	–
D	Power Economics and Regulatory Mechanism	–	–	–	✓	✓	–	–	–
E	Renewable and Alternative Energy Sources	✓	✓	✓	✓	–	–	–	–
E	Renewable Energy Project Analysis	–	–	✓	–	–	–	–	–
E	Renewable Energy Project Development and Finance	–	–	–	–	–	✓	✓	✓
E	Renewable Energy Sources	–	–	–	–	✓	✓	✓	✓
F	Sectoral Studies in Energy and Environment	–	–	✓	✓	✓	✓	–	–

The programme at SIIB also used the climate centrality approach to frame courses as per international market driven scenarios and recent industry and regulatory driven approaches in curriculum development. Such innovative market based mechanisms which were first created back in the mid-2000s starting from the UNFCCC driven Clean Development Mechanism process therefore became an integral part of the curriculum in the early years of the programme (Table 4.3). The business relevance of the course was linked to the fact that several new project developers wanted to invest in sustainable development projects which were lucrative in terms of return on investment. Courses on CDM and Project Design Document methodology became a necessity for students at SIIB during 2009–12. This also helped industry players including small and medium enterprises (SMEs) (Saini et al., 2012). As new market mechanisms evolved both internationally (Bali Action Plan of 2007) and in India due to emergence of policies at National level like NAPCC of 2008, the Institute brought in elements of the industry trends and policy changes through curricular inputs. The recently established PAT scheme (Perform, Achieve and Trade) of the Government of India through the Bureau of Energy Efficiency is one such step which is likely to provide a major impetus to strengthening energy efficiency practices. Similarly, a course on the concept of Renewable Energy Certificates (REC) was also created to involve industry players at providing market based incentives to project developers in the renewable energy space in increasing the overall share of renewable energy in the overall energy mix of India.

4.4.2 Power Economics, Governance, and Energy Markets

The evolution of the electricity sector and subsequent reforms in India has brought about significant changes in power sector reforms in the country (Prayas, 2017). Following the Electricity Act, 2003 of GoI, an attempt has been made to mandate the development of a transparent, fair and equitable process in the generation, transmission and distribution of electricity across the country. The clear link between electricity use – energy growth – GHG emissions and climate change in industries has led to several private sector entities trying to capture a market share of the power sector business. The existing programme has tried to include these issues as part of a holistic curriculum (Table 4.4) which brought in course elements using climate change as a keystone course around areas like energy efficiency, technology, regulatory issues, consumer needs, and social equity.

Table 4.3 Climate Change Specific Courses at SIIB

Courses	2009–11	2010–12	2011–13	2012–14	2013–15	2014–16	2015–17	2016–18
Carbon Financing and Project Design Document	–	–	√	√	–	–	–	–
Carbon Financing and Market Instruments	–	–	–	–	√	–	–	–
Climate Change	√	√	–	–	–	–	–	–
Climate Change Impacts, Mitigation & Adaptation	–	–	√	–	√	–	–	–
Climate Change, Carbon Markets and Financing	–	–	–	–	–	√	√	√
Understanding Carbon credits	–	–	–	√	√	–	–	–
Understanding Carbon Credits and Markets	–	√	√	–	–	–	–	–
Renewable and Alternative Energy Sources	√	√	√	√	–	–	–	–
Renewable Energy Project Analysis	–	–	√	–	–	–	–	–
Renewable Energy Project Development and Finance	–	–	–	–	√	√	√	√
Renewable Energy Sources	–	–	–	–	√	√	√	√

Table 4.4 Energy and Power Courses at SIIB

S. No.	Courses	2009–11	2010–12	2011–13	2012–14	2013–15	2014–16	2015–17	2016–18
B	Energy Asset Management	√	–	–	–	–	–	–	–
B	Energy Audits	√	–	–	–	–	–	–	–
B	Energy Audits and Energy Conservation	–	√	–	–	–	–	–	–
B	Energy Audits & Energy Management	–	–	√	√	√	√	√	√
B	Energy Conservation	√	–	–	–	–	–	–	–
B	Energy Conservation & Sustainable Energy Development	√	–	–	–	–	–	–	–
B	Introduction to Energy Regulation and Policies	–	√	√	√	√	√	√	√
B	Non-Renewable Energy Sources	√	√	√	√	√	√	√	√
C	Contracting in Energy Sector	–	–	–	–	–	√	√	√
C	Energy Derivatives and Risk Management	–	–	√	–	–	–	–	–
C	Energy Economics	√	√	–	–	–	–	–	–
C	Energy Marketing/Power Trading	√	√	–	–	–	–	–	–
C	Energy Modeling	√	√	–	–	–	–	–	–
C	Energy Project Appraisal & Financing	–	–	–	√	√	–	–	–
C	Financing the Energy Sector	√	–	–	–	–	–	–	–
C	Power Trading	–	–	√	–	–	–	–	–
C	Power Trading & Energy Markets	–	–	–	–	√	√	–	–
C	Project Financing in Energy Sector	–	√	–	–	–	–	–	–
D	Electricity Regulatory Issues and Mechanisms	–	–	–	√	–	√	√	√
D	Power Economics and Regulation	–	–	√	–	–	–	–	–
D	Power Economics and Regulatory Mechanism	–	–	–	√	√	–	–	–

4.5 Conclusion

The significance of climate change continues to dominate global politics and remains one of the most important global environmental challenges (Pachauri and Resinger, 2007) faced by mankind. There is very little time for the world to take action in combating impacts of GHG emissions which are likely to increase in an exponential way (Meinhausen et al., 2009). The rising demand for energy and its consumption in order to achieve higher economic growth is cited as a key driver of higher GHG emission rates (IEA, 2008). Building local capacity on such International issues like climate change through business education will therefore be largely required at different academic institutions to develop contemporary knowledge and skill sets of young managers who opt of a career in the field of emerging area of sustainability. In order to help and build sustainable and low carbon economy, strategic involvement of not only the country governments is needed but also business and industry, academic institutions and civil society needs to work in close cooperation and forge meaningful partnerships for implementation.

References

Adams, C. A., Heijltjes, M. G., Jack, G., Marjoribanks, T., and Powell, M. (2011). The development of leaders able to respond to climate change and sustainability challenges: the role of business schools. *Sustain. Account. Manag. Policy J.* 2, 165–171.

Agarwal, A., and Narain, S. (1991). *Global Warming in an Unequal World: A Case of Environmental Colonialism.* New Delhi: Centre for Science and environment.

Amran, A., Nabiha Abdul Khalid, S., Abdul Razak, D., and Haron, H. (2010). Development of MBA with specialisation in sustainable development: the experience of Universiti Sains Malaysia. *Int. J. Sustain. High. Educ.* 11, 260–273.

Bali Action Plan (2007). Decision-/CP. 13. [accessed January, 12, 2012].

Battisti, D. S., and Naylor, R. L. (2009). Historical warnings of food insecurity with unprecedented seasonal heat. *Science* 323, 240–244.

Byravan, S., ChellaRajan S., and Rangarajan R. (2012). Sea level rise – Impact on major infrastructure and land along the Tamil Nadu coast. In Handbook of Climate Change and India – development, politics and governance. Ed. N Dubash, EarthScan UK pp. 41–50.

Carbon Trust (2006). *Higher Education Carbon Management Programme.* Available at: http://www.carbontrust.co.uk/carbon/he/ November 2006 [Accessed February 15, 2017].

Chambers, D. (2009). Assessing & planning for environmental sustainability: a framework for institutions of higher education. *Sustainability at Universities: Opportunities, Challenges and Trends*, Series: Umweltbildung, Umweltkommunikation und Nachhaltigkeit/Environmental Education, Communication and Sustainability, *WL Filho, Franfurt* pp. 287–297.

Coops, N. C., Marcus, J., Construt, I., Frank, E., Kellett, R., Mazzi, E., and Schultz, A. (2015). How an entry-level, interdisciplinary sustainability course revealed the benefits and challenges of a university-wide initiative for sustainability education. *Int. J. Sustain. High. Educ.* 16, 729–747.

Dasgupta, C. (2012). "Present at the creation: the making of the UN framework convention on climate change," in *Handbook of Climate Change and India – Development, Politics and Governance*, ed. N. Dubash (London: Earthscan), 89–97.

Date-Huxtable, E., Ellem, G., and Roberts, T. (2013). "The low carbon curriculum at the University of Newcastle, Australia," in *Sustainability Assessment Tools in Higher Education Institutions – Mapping Trends and Good Practices Around the World,* eds. S. Caeiro, W. Leal Filho, C. Jabbour, U. M. Azeiteiro (Berlin: Springer) Part IV, 345–357.

Godemann, J., Herzig, C., Moon, J., and Powell, A. (2011). Integrating sustainability into business schools—Analysis of 100 UN PRME Sharing Information on Progress (SIP) reports. ICCSR Research paper 58-2011. *Nottingham: International Centre for Corporate Social Responsibility*, University of Nottingham, Nottingham (58-2011).

IEA (2007). *World Energy Outlook: China and India Insights.* International Energy Agency, Paris.

IPCC (2007). Summary for policymakers. Solomon, S., Qin, D., Manning, M., Chen, Z., Marquis, M., Averyt, K. B., & Miller, H. L. (2007). IPCC, 2007, *Climate change.* p. 79.

IPCC (2014). Summary for Policymakers. In: Climate Change 2014: Mitigation of Climate Change. Contribution of Working Group III to the Fifth Assessment Report of the Intergovernmental Panel on Climate Change [Edenhofer, O., R. Pichs-Madruga, Y. Sokona, E. Farahani, S. Kadner, K. Seyboth, A. Adler, I. Baum, S. Brunner, P. Eickemeier, B. Kriemann, J. Savolainen, S. Schlömer, C. von Stechow, T. Zwickel and J. C. Minx (eds.)]. Cambridge University Press, Cambridge, United Kingdom and New York, NY, USA.

IPCC (2014). Climate Change 2014: Synthesis Report. Contribution of Working Groups I, II and III to the Fifth Assessment Report of the Intergovernmental Panel on Climate Change [Core Writing Team, R. K. Pachauri and L. A. Meyer (eds.)]. IPCC, Geneva, Switzerland, 151 pp.

Jakobsen, S (1998). India's position on climate change from Rio to Kyoto: a policy analysis, *CDR working paper 98.11,Centre for Development Research, Copenhagen.*

Jindal, K. K., and Mankotia, M. S. (2004). Impact of changing climatic conditions on chilling units, physiological attributes, and productivity of apple in western Himalayas. *Acta Hort.* 662, 111–117.

Leal Filho, W. (2011). About the role of universities and their contribution to sustainable development. *High. Educ. Policy,* 24, 427–438.

Leal Filho, W. (ed.) (2012) Sustainable Development at Universities: New Horizons Series: Umweltbildung, Umweltkommunikation und Nachhaltigkeit/Environmental Education, Communication and Sustainability. Vol. 34, pp. 994. Frankfurt, Peter Lang Scientific Publishers.

Lovejoy, T., and Hannah. L. J. (Eds.) (2006). *Climate Change and Biodivesity.* Yale University Press, New Haven.

Meinhausen, M., Meinhausen, N., Hare, W., Raper, S. C. B., Frieler, K., Knutti, R., et al. (2009). Green house gas emission targets for limiting global warming to 2°C". *Nature* 458, 1158–1163.

Ministry of Environment and Forests (MoEF), India (2010). India: greenhouse gas emissions, 2007.

Muller, B., Hohne, N., and Ellermann, C. (2009). Differentiating historic responsibilities for climate change. *Clim. Policy* 9, 593–611.

Naeem, M., and Neal, M. (2012). Sustainability in business education in the Asia Pacific region: a snapshot of the situation. *Int. J. Sustain. High. Educ.* 13, 60–71.

Pachauri, R. K., and Reisinger, A. (2007). *Contribution of Working Groups I, II and III to the Fourth Assessment Report of the Intergovernmental Panel on Climate Change.* Core Writing Team, IPCC, Geneva, Switzerland. pp 104.

Pandey D., Agrawal, M., and Pandey, J. S. (2011). Carbon footprint: current methods of estimation. *Environ. Monit. Assess.* 178, 135–160.

Parikh, J. (1995). North–South cooperation in climate change through joint implementation. *Int. Environ. Aff.* 7, 22–43.

Patil, Y., and Rao, P. (2014) "Industrial waste management in the era of climate change - a smart sustainable model based on utilization of active and passive biomass," in *Handbook on Climate Change Adaptation,* ed. W. Leal Filho, (Berlin: Springer), 2079–2092.

Prayas Energy Group (2017). Many sparks but little light: The rhetoric and practice of electricity sector reforms in India. In: *Prayas Energy Group*, pp. 389.

Pulver, S. (2002). Organising business Industry NGOs in the climate debates, *Greener Management International*, Vol. 39, Autumn, pp. 55–67.

Rajamani, L. (2007). India's negotiating position on climate change: Legitimate but not sagacious, CPR Issue brief, No. 2, Centre for Policy Research, New Delhi.

Rao Prakash, Patil Yogesh, Gupte Rajani (2013). Education for Sustainable Development: Trends in Indian Business Schools and Universities in a Post Liberalization Era. In: *Sustainability Assessment Tools in Higher Education Institutions – Mapping Trends and Good Practices Around the World* (Ed. Sandra Caeiro, Walter Leal Filho, Charbel Jabbour, Ulisses M. Azeiteiro), Springer International Publishing, Switzerland, Part IV, pp. 417–432.

Rao, P. (2011). "Integrating sustainability into global business practices-an emerging tool for management education," in *Internationalisation of Higher Education* eds. Rajani Gupte, B. Venkataramani, and D. Gupta), pp. 60–73. Excel Publishers, India.

Rao, P. and Patil, Y. (2015). "Integrating energy and environment in postgraduate academic management education: a case study," in *Implementing Campus Greening Initiatives, Approaches, Methods and Perspectives*, (Berlin: Springer), 265–276.

Roy, R., Potter, S., and Yarrow, K. (2008). Designing low carbon higher education systems: environmental impacts of campus and distance learning systems. *Int. J. Sustain. High. Educ.* 9, 116–130.

Saini, S., Rao, P., Patil, Y. (2012). City based analysis of MSW to energy generation in India, calculation of state-wise potential and tariff comparison with EU. *Procedia Soc. Behav. Sci.* 37, 407–416.

Sant, G., and Gambhir, A. (2012). "Energy development and climate change," in *Handbook of Climate Change and India – Development, Politics and Governance*, ed. N. Dubash (London: Earthscan) 289–302.

Sengupta, S. (2012). "International climate negotiations and India's role," in *Handbook of Climate Change and India – Development, Politics and Governance*, ed. N. Dubash (London: Earthscan publications), 101–117.

Shriberg, M. (2012). "Building sustainability leaders: a framework to prepare students to thrive on complexity and lead transformative changes," in *Environmental Education, Communication and Sustainability, Sustainable Development at Universities—New Horizons*, ed. W. Leal Filho (Bern: Peter Lang), 19–28.

(unpublished) Spranger, S. (2011), *"Calculating the carbon footprint of Universities"*. Masters Thesis, Economics and Informatics, Erasmus School of Economics. Unpublished Thesis, pp. 107.

Sreekumar, N., and Dixit, S. (2010). *Electricity for All: Ten Ideas Towards Turning Rhetoric Into Reality*. Prayas energy Publication. Available at: http://www.Prayaspune.org/peg/Publications/item/84.html [accessed May 2, 2011].

Stubbs, W. (2013). Addressing the business-sustainability nexus in postgraduate education. *Int. J. Sustain. High. Educ.* 14, 25–41

Stubbs, W. and Cocklin, C. (2008). Teaching sustainability to business students: shifting mindsets. *Int. J. Sustain. High. Educ.* 9, 206–221.

UNCSD (2012). "Current ideas on sustainable development goals and indicators," *Proceedings of the United Nations Conference on Sustainable Development (UNCSD), Secretariat (No. 6)*. New York.

UNFCCC, U. (1992). Framework Convention on Climate Change. *Palais des Nations, Geneva*.

University of Toronto. (2010), St. George Campus Greenhouse Gas Emissions Inventory Report.

Viebahn, P. (2002). An environmental management model for universities: from environmental guidelines to staff involvement. *J. Clean. Prod.* 10, 3–12.

WWF (2005). *An Overview of Glaciers, Glacier Retreat, and Subsequent Impacts in Nepal, India and China*. Washington, DC: WWF Nepal Publication.

5

Linking Curriculum Innovation and Campus Greening in the Transition to a Low-Carbon Economy: A Case Study

Judy Rogers[1] and Karolina Bartkowicz[2]

[1]School of Architecture and Design, RMIT University,
Melbourne, VIC, Australia
[2]School of Property, Construction and Project Management,
RMIT University, Melbourne, VIC, Australia

Abstract

Transitioning to a low carbon economy (LCE) represents one of the most significant and urgent challenges facing the international community. The need for such a transition has gathered momentum since the Paris Agreement in late 2015 where 195 nations agreed to a binding climate change deal aimed at limiting global warming to a 2°C rise. This is a complex challenge, particularly for Universities as the key centres for innovation globally. Mitigation, it is argued here, involves more than campus greening initiatives and education about climate change and reduction of CO_2 emissions, it involves instead a complex process of encouraging behavioural change to address consumption patterns and waste while at the same time providing increasing opportunities for participation and innovation. This chapter addressed one small aspect of this challenge highlighting the importance of integrating campus greening strategies with hands-on experiential project-based learning and teaching initiatives for students and staff, as an approach to pursue climate change mitigation and support Higher Education Institutions (HEIs) through the transition to a LCE. The chapter discusses, through a multi-stage green roof pilot project delivered at RMIT University, 2017, methods for harnessing opportunities and addressing barriers to green roof uptake. The project resulted in institutional learning and change leading to further investment and implementation of permanent green roof projects that will provide research and educational opportunities into the future. Project-based, experiential learning it is argued here can, therefore, lead to multiple benefits for learners, for teachers and for institutions in the transition to a LCE.

5.1 Introduction

Addressing human-induced climate change is one of the most significant challenges facing the international 'community', requiring a transformational agenda (OECD/IEA/NEA/ITF, 2015). As emissions of greenhouse gases continue to rise, the urgency increases. Impacts on food security, biodiversity, oceans, land, and water resources, along with the impacts of extreme weather events pose immediate and ever increasing threats. The Intergovernmental Panel on Climate Change [IPCC], 2014 *Fifth Assessment Report* (2014) identified that recent anthropogenic emissions of greenhouse gases are the highest in history and that human influence is "extremely likely" to be the main cause. The Report identified multiple mitigation pathways that are likely to limit warming to below 2°C relative to pre-industrial levels. However, these pathways would require substantial emissions reductions over the coming decades and near zero emissions of CO_2 by the end of the century (IPCC 2014). Implementing strategies to address these reductions towards the transition to a low or zero carbon economy poses significant technological, economic, social, and institutional challenges (IPCC 2014).

The need for such a transition has gathered momentum since the Paris Agreement in late 2015 where 195 nations agreed to a binding climate change deal aimed at limiting global warming to a 2°C rise. Despite the urgency, however, the World Economic Forum's 11th *Global Risks Report 2016* identified the failure of climate change mitigation and adaptation as 'the most impactful risk' for the years to come, ahead of weapons of mass destruction, ranking 2nd, and water crises, ranking 3rd (World Economic Forum [WEF], 2016). Likewise, the Agenda for Sustainable Development 2030 (United Nations 2015) noted with

>...*grave concern the significant gap between the aggregate effect of parties' mitigation pledges in terms of global annual emissions of greenhouse gases by 2020 and aggregate emission pathways consistent with having a likely chance of holding the increase in global average temperature below 2°C or 1.5°C above pre industrial levels.*

Protecting 'the earth' and the people who inhabit it from further 'degradation' through sustainable consumption and production, managing natural resources and addressing climate change are among the goals identified in *Transforming our world: the 2030 Agenda for Sustainable Development*. The Agenda identified 17 'integrated and indivisible' goals along with 169 targets for the achievement of sustainable development by 2030 which demand that reducing carbon emissions and addressing climate change cannot be separated out as singular 'issues' from other concerns such as reducing poverty (goal 1) and inequality (goal 10), gender equity (goal 5), access to reliable energy (goal 7) and education (goal 4), inclusion (goal 16) and partnerships (goal 17) amongst others.

These are complex challenges, particularly for Universities as key centres for innovation, education, and research globally. However, as the World Economic Forum's (2016) report argued

> *The increasing volatility, complexity and ambiguity of the world not only heightens uncertainty around the "which", "when", "where" and "who" of addressing global risks, but also clouds the solutions space. We need clear thinking about new levers that will enable a wide range of stakeholders to jointly address global risks, which cannot be dealt with in a centralized way.*

Here, the emphasis is on the 'new' – new ways of thinking and doing that move beyond the siloed disciplinary approaches that have historically and continue to characterise Higher Education Institutions (HEIs) across the globe along with the challenge to reconsider what constitutes a 'solution' or 'solutions' in a complex, ambiguous and uncertain world when 'we' do not in fact know what the 'problem' is and where problem(s) identification requires new and as yet unthought of modes of thinking. And so, as the OECD/IEA/NEA/ITF (2015) has argued

> *The case of climate change also highlights the need for new inter-disciplinary and trans-disciplinary research environments. The diverse sources of knowledge upon which "climate" innovation, and environmental innovation more generally, draws makes this particularly important in the context of policy alignment for the low-carbon transition.*

Project based experiential learning has increasingly been identified as central to education for sustainability (Brundier and Wiek, 2013; Leal et al., 2016;

Wiek et al., 2014) with its emphasis on building the capacity of students as change agents. This chapter reports on the outcomes of one such learning experience at RMIT University, Melbourne, Australia which provided opportunities for students to be exposed to and engage with the challenges and opportunities involved in implementing measures to reduce the impacts of climate change. Rather than teaching students about the need for a transition to a low-carbon economy (LCE), students learnt by doing.

5.2 The Role of HEIs in the Transition towards Sustainability

As Leal et al. (2016) argue 'Universities should act as agents of change through promoting the principles of sustainable development within their institutions and in society'. This requires a move away from compartmentalised, disciplinary-based learning towards an approach that allows for innovation, risk taking, and cross-disciplinary engagement. University campuses can then provide fertile ground to not only reduce energy consumption and waste, encourage behavioural change, but also act as laboratories for researching and testing through curriculum innovation. In this scenario, University operations, research, learning, and teaching cannot be viewed separately but rather become a platform for developing and testing new ideas. As Leal et al. (2016) suggest

> *Learning through experience, in real world scenarios, enables the development of competences that will not only be valued in the workplace but will be vital, if the world is to rectify unsustainable development. Such projects will enhance sustainability literacy and when deployed on campus, are likely to also contribute to institutional learning and more integrative approaches to sustainability'.*
> (pp. 126–135)

Within the learning and teaching space, this demands a pedagogy that shifts the role of the teacher away from 'expert' to that of facilitator. As Dobson and Tomkinson (2012) suggest '*Rather than learning from an expert, the students become the "experts" in the scenario that they are investigating, role-playing a professional consultancy team. This encourages creative thinking by students instead of simply seeking to replicate a "model" answer.* Or, as Davis and Sumura (2007) suggest 'what teaching is can never be reduced to or understood in terms of what the teacher does or intends.

Rather, teaching must be understood in terms of its complex contributions to new, as-yet-unimaginable collective possibilities'. They argued

Teaching here is more about a conscientious participation in expanding the space of the possible by creating the conditions for the emergence of the not-yet-imaginable, rather than about perpetuating entrenched habits of interpretation (or even exploring the limits of current imagination). Teaching, like learning, is not about convergence onto a pre-established truth, but about divergence – about broadening what can be known and done. In other words, the emphasis is not on what is, but on what might be brought forth. Teaching thus comes to be a participation in a recursively elaborative process of opening up new spaces of possibility while exploring current spaces. (pp. 263–278)

This presents a range of challenges for HEIs because as Leal et al. (2016) argue in '*the main, the curriculum across most universities is developed to provide students with an increasingly narrow understanding of courses, professions and jobs, with a focus on specific knowledge and skills. A curriculum embracing education for sustainability requires a broader approach than just discipline knowledge.*' Education for sustainable development is more about process and less about content where learners have the opportunity to engage with and challenge different ways of knowing, doing and thinking. Sustainability education must also equip students with '*interpersonal competencies and transdisciplinary/transacademic work experience. These skills cannot be developed through lecture-based activities alone, but require hands-on practice, teamwork and community engagement opportunities*' (Wiek et al., 2014).

This chapter reports on the development of one such hands-on learning space at RMIT University's Melbourne Campus. The focus was on the design, construction and monitoring of three green roof pilot projects on RMIT city campus, Melbourne, Australia. The pilot projects opened up a discursive space for staff, students, and property services to share, challenge, evaluate and test how urban campus greening strategies might lead to further innovation in mitigation and adaptation. The project identified methods for harnessing opportunities and addressing barriers to green roof uptake, resulting in institutional change and further investment and implementation of permanent green roof projects that will provide further research and educational opportunities into the future. Beyond the knowledge and skills gained in design and construction, the emphasis on group work in the course allowed a space to open for collaboration and 'affective' learning (Shephard, 2008). Working together to produce an outcome developed skills in negotiation, listening, responding, and learning to value differences in perspective. Also having the opportunity to monitor and evaluate the green roof modules assisted students to move beyond the easy assumptions often made about sustainability in the urban context, to think critically about the way in which the 'sustainable city' is performative and how, given this understanding, they could in fact contribute to new, as-yet-unimaginable collective possibilities' (Davis and Sumura, 2007).

The chapter acknowledges that urban greening strategies, or even green roofs, in and of themselves will not lead to a LCE nor will they address the complex issues that emerge in a changing climate. They exist as one small approach that opens up a space for other spaces to emerge and to rethink what a University campus in the future could be. The project sits within a suite of projects currently being rolled out across RMIT University campuses. The Sustainable Urban Precincts Program (SUPP), for instance, aims to reduce electricity use over eight years by an estimated 263 million kilowatt hours, leading to a 32,000-tonne reduction in greenhouse gas emissions. Water use will be cut by an estimated 53 million litres per year (RMIT, 2017). Each stage of the project discussed here also informed RMIT's New Academic Street project (NAS) contributing to its sustainability benchmarks. The NAS project is a construction program that aims to upgrade facilities including the Swanston library, the Student Hall, collaboration spaces, and food options and transform the heart of the RMIT City campus (RMIT, 2017). NAS aims to

> *provide a mechanism for student participation, creating a living laboratory. There are many aspects of a building such as view, access to daylight, lighting and air quality which contributes to good Indoor Environment Quality (IEQ). Good IEQ has been linked to improve productivity, wellbeing and occupant satisfaction...Also within the new NAS rooftops there will be growing space allocated for students in programs such as Landscape Architecture to test soils and plants that work well on urban green roofs."*
> *(RMIT, 2017)*

A key feature of the NAS project includes opening the campus to the surrounding streetscape and creating light-filled laneways, glass-roofed arcades and rooftop urban spaces (RMIT, 2017). The University's vision is that all campuses should contribute to urban sustainability demonstrating leadership in sustainable design and innovation. A key component of the project reported here is that while it informed the broader university strategy it also offered opportunities for student involvement and participation. Planting schemes suitable for rooftops as well as lightweight growth medium blends were trialled via the Matter of Landscape project, providing detailed information to the NAS project team, resulting in suitable choices for planting and growth medium palettes incorporated within the NAS project. Rooftop spatial design, module depth and layout as well as the Post Occupancy Evaluations provided an insight into the infrastructure required for a learning laboratory to facilitate outdoor collaboration and learning spaces for students, researchers, and industry professionals.

5.3 The Matter of Landscape: Sustainable Design Strategies for RMIT City Campus

Along with green walls, rain gardens, street trees, and permeable paving, green roofs are part of an emerging global discourse around city 'greening'. 'Green roof' is an umbrella term used to describe a number of systems for green rooftops. The claimed benefits of green roofs reflect outcomes across the mitigation/adaptation spectrum. The list of benefits is broad and widely shared. They include reduction in the heat island effect, noise reduction, energy conservation, amenity, increased biodiversity, habitat for wildlife, replacement of lost green space, increased property values, storm-water management along with what has been described as 'green relief' in highly dense cities. Green roofs it is further claimed transform cities from urban grey to green while at the same time mitigating the effects of climate change. Barriers to uptake of green roofs are most often identified as a lack of information, a lack of awareness, overall cost, and the lack of immediate economic return (Claus and Rousseau, 2012; Wilkinson and Reed, 2009) and so a range of policy measures have been developed or have been proposed within the Australian context, and globally, to provide incentives for uptake. The overriding focus within this policy discourse is on market based incentives— encouraging private investment for public environmental 'good'—along with an emphasis on individual building performance using efficiency metrics and rating tools. Within this context little distinction is made between different types of green roofs: the assumption being that all green roofs deliver similar sustainability benefits.

In order to explore and to evaluate these claimed benefits, the *Matter of Landscape* project began in 2012 with a series of comparative precedent and best practice case studies in Melbourne of existing green roofs, background research on various typologies of green roofs, soil or growth mediums and plant materials. From these initial studies, the pilot green roof modules were developed and suitable vegetation (testing biodiversity, edible landscapes, and carpeting species for thermal insulation and fast growth) as well as alternative growth mediums (testing light weight and soil mixtures). The project was then taken into a living laboratory contexts, where third year students built, maintained, and monitored the performance of a series of green roof mobile modules in building 8 on a level 10 balcony (Figure 5.1). Part 2 of the project in 2014 tested planting modules at a larger scale, focusing in particular on biodiversity plantings, using larger tree species, shrubs, and ground covers. third year students also experimented with and designed more robust cladding systems using recycled materials. A vegetable plot was also planted and maintained by RMIT student Union with produce being used in the Real Foods Café. In addition, a post-occupancy evaluation was undertaken in semester 2 that involved comparing the green roof pilot project and a public space at ground level to identify social use of the spaces along with environmental monitoring of the green roof modules. Case study research into best practice sustainable design was also undertaken in Singapore, focusing in particular on social sustainability outcomes. In 2015, the project shifted focus to a level 8 balcony where students once again designed and constructed green roof modules. Both intensive and extensive modules were planted with a focus on

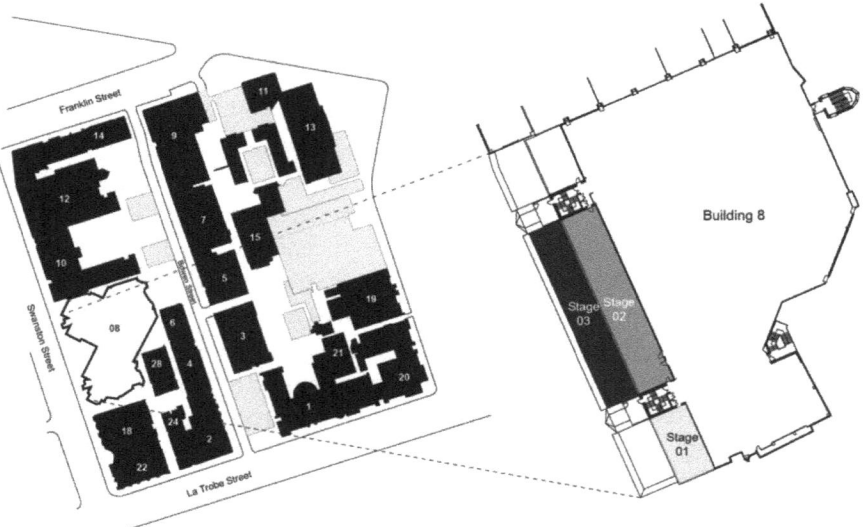

Figure 5.1   RMIT University City campus map and pilot green roof site locations.

biodiversity plantings. Students also designed seating for the balcony as this site has the potential for greater levels of access. Monitoring in semester 2, 2015 also included plant performance, growth medium performance, and a post-occupancy evaluation. During this time, monitoring also continued on the level 10 balcony. Plant monitoring on both the level 8 and level 10 balconies generated a great deal of rich data on plant and growth medium performance.

5.4 Learning by Doing

The Matter of Landscape Project provided students with practical learning opportunities that not only enhanced their capabilities and connected their education to real world issues, but also delivered methods to quantify and qualify the effects of their design and construction work. Through a project-based learning approach, students were introduced to core sustainability concepts and research methodologies. The project sought to combine learning to design and construct a roof garden and to measure its tangible effects. The project was a multistage one that developed over 4 years with each stage building on what went before.

5.4.1 The Matter of Landscape: Stage 1

The initial green roof experimental project was located on a south-westerly balcony on the tenth floor of RMIT University City Campus. This particular site was chosen for its predominant harsh wind conditions and minimal exposure to sunlight. The individual test plots consist of 16 elevated mobile planters (modules) that were assembled to accommodate trials of extensive and intensive planting modes (Figure 5.2). These comprised of nine deep planters (600 mm in depth) and six shallow planters (180 mm in depth) providing a range of growing conditions according to their positioning. One deep planter was included as a control without any plant material.

With the advice of industry professionals, students proposed planting schemes and growth medium mixes. The plantings included: biodiversity plots, edible landscapes plots, and carpeting species plots (Figure 5.3). Students collected performance-based data on plant growth (including survival rates, vegetation coverage, plant proliferation rates, physical damage, sun scorch, and evidence of reaction to the watering regime), soil performance (porosity, drainage, hydration, and electrical conductivity) as well as albedo effect (reflective surface temperatures) and thermal insulation performance (Figure 5.4).

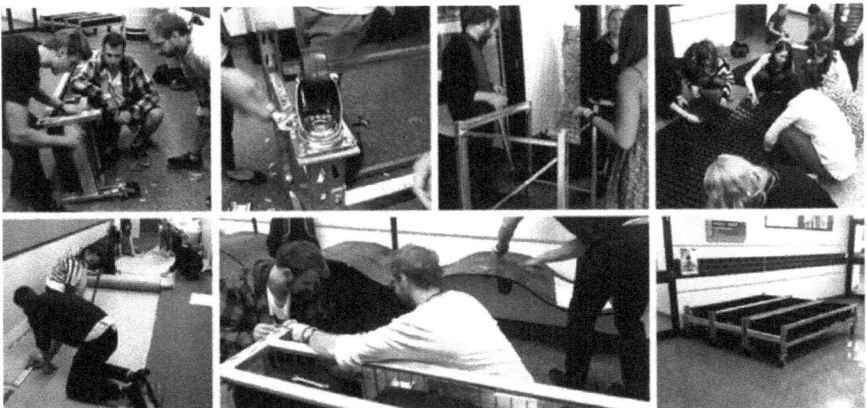

Figure 5.2   Landscape Architecture students assembling green roof modules.

Figure 5.3   Landscape Architecture students planting the green roof modules.

Experimental cladding systems were also installed and offered a range of benefits. The cladding provided an insulation barrier to the growth medium containers, opportunities for seating and catchment for storm water and self-watering opportunities. Some cladding systems also encompassed lighting to provide visibility during night hours.

Figure 5.4 Environmental monitoring by landscape architecture students.

In addition, to quantitative environmental metrics the students investigated various qualitative or more socially sustainable outcomes. Because the roof top space can be viewed from offices above, it was important to design it as a place for visual respite. This prompted students to design and test various bench and seating prototypes using light-weight and recycled material, to allow visitors to spend time on the balcony. This had an added effect of providing further insulation to module growing mediums. Other environmental aspects such as water collection, drainage, wind protection, and enhancement of natural sunlight were integrated within module cladding and seating systems.

The momentum of the project also facilitated wider university involvement including interdisciplinary collaborations between RMIT's Environmental Science Program and Property Construction and Management Program. Beyond the student learning aspects of the project, a final research report was developed for RMIT Property Services as a briefing document, case study comparison, and initial design guidelines for the commissioning future roof gardens.

5.4.2 The Matter of Landscape: Stage 2

Stage two of the project began at the beginning of 2014 with the aim of building on outcomes from stage 1 through further student-led environmental monitoring including the development and testing of techniques to evaluate biodiversity and productive landscape benefits, as well as microclimatic condition. Here, students developed their own experiments based on literature, previous experiments and their own observations regarding prevailing conditions on site. A new site had been selected for the experiment that allowed for increased access for University staff and student in order to investigate how such a space could be used. The parameters of this new experiment responded to the data collected in 2013 and 2014 and included a post-occupancy evaluation to address social sustainability criteria in greater detail. The focus in stage 2 of the project was also to encourage students to look beyond the 'easy' and often unsubstantiated assumptions that all green roofs and green infrastructure are 'sustainable' through investigation on a broad scale through to the 1:1 construction of a pilot project green roof. The project began, therefore, with a provocation where students were asked—Green Roofs and Green Infrastructure—are they sustainable? Responding to this question involved in the first instance an evaluation of an existing green roof using sustainability principles as a framework. The evaluation framework reflected claimed sustainability benefits drawn from broader literature and students were asked to seek out available evidence that supported the claimed benefit.

The students were also asked in the initial stages of the semester to divide into groups of five to develop a master plan for RMIT city plan identifying potential sites for green roofs. Each group was asked to explicitly focus on either the social, economic, or environmental dimensions of sustainability in their plans. However, by week 2 what they all very soon realised is that it is not possible to separate each of these dimensions out so easily. For instance, students working on environmental sustainability quickly realised that they needed to engage with the social and economic dimensions of sustainability in order to realise their ambition for RMIT city campus. What is interesting here is that students could have been told that each of the sustainability dimensions needed to be integrated but rather they learnt this though the process of negotiating their master plan (Figure 5.5). This was reflected in the process and methodology the students used to generate their group master plans, combining and integrating various elements of the three sustainability pillars to realise their key ambitions through development of specific roof-top typologies applied to the RMIT City Campus. In addition, realising the need to engage with all three dimensions of sustainability all of the groups identified that in order to engage with the social realm the question of access needed to be foregrounded as a key challenge.

Students then constructed green roof modules for the new site, this time testing planting modules at a larger scale with tree species, further testing of light-weight soil/growth mediums, and identification of species appropriate for supporting biodiversity and edible landscapes. Students also tested recycled materials and a more robust cladding system and develop ongoing maintenance regimes.

Figure 5.5 Negotiating and discussing a master plan for RMIT city campus.

Each of the four tutorial groups were then divided into two with one team responsible for the hard landscape component of the green roof modules. This included: working to an allocated budget, designing, documenting, and constructing a cladding system to be fixed to one or more of the allocated green roof modules. The other team was responsible for the soft landscape component and will involve developing a planting plan, documenting and then implementing a planting scheme within the allocated green roof modules. These activities were followed in semester 2, again with Landscape Architecture Students, with further environmental monitoring and the development of, and testing of a Post-occupancy evaluation with a particular focus on how the site is used.

5.4.3 The Matter of Landscape: Stage 3

With a further focus on biodiversity, stage 3 of the project involved further comparative case studies resulting in an understanding of how emerging 'green technologies' could influence the design and construction of the third pilot green roof project on the level 8 balcony of building 8, RMIT City Campus.

The design of the Level 8 rooftop was also a result of a green roof competition where students were invited to respond to a set brief incorporating 4 biodiversity themed plots (The Urban Meadow, The Rocky Mountains,

The Urban Grassland, and The Succulent Garden), an urban agricultural plot that would be cared for by the RMIT Student Union (Greening RMIT) and build upon current site challenges and behaviours (environmental and social).

The level 8 rooftop is 31 m in length and 5 m in width (forming part of the roof to level 7) and located adjacent to the level 10 rooftop, separated by a 6-m elevation. The rooftop previously functioned as an event space and an outdoor study area, however, was inaccessible to the building occupants prior to the implementation of the pilot green roof. There is visual access to the site via adjacent staff offices and study area on level 8 which face directly onto the rooftop and the space can be seen from a number of surrounding buildings—including RMIT's Swanston Academic Building. Similar to the level 10 rooftop, the level 8 rooftop suffers from harsh wind conditions and has nominal sun exposure. The level 8 rooftop, experiences different micro climatic conditions from the level 10 rooftop as it has an additional balustrade structure providing further shadowing across the rooftop, less sun exposure due to the lower height above ground level and is the storm water outlet from the level 10 rooftop – therefore conditions are cooler and wetter than the level 10 rooftop. This particular rooftop also differs from the level 10 rooftop as it had student and staff access (opened in semester 2, 2015). The level 8 balcony location was chosen as a suitable space for creating a living laboratory due to its location within the CBD context and site features—appropriate parapet height, drainage elements, structural loading, etc. These conditions provided the project with optimal research opportunities for testing, promoting and evaluating biodiversity within a green roof context.

The competition called for an innovative spatial design solution (Masterplan) for the Building 8, Level 8 Balcony Rooftop, using 3 styles of elevated green roof modules: extensive (shallow planting), intensive (deeper shrubbery planting), and giant (deepest – tree planting), to transform the rooftop form an unused and abandoned balcony into the third pilot green roof demonstration project (Figures 5.6 and 5.7).

Stage 3 of the project continued to build upon the achievements of part 1 and 2 linking university operations, learning and teaching outcomes, research development, and industry engagement via establishment of industry partners—Live Roofs Australasia and Urban Commons. Collaboration with the RMIT Student Union and the Student Landscape Architecture Body (SLAB) meant that students were actively involved in the ongoing rooftop monitoring and evaluation as well as maintenance.

Figure 5.6   Students preparing the green roof modules on the level 8 balcony.

Figure 5.7   Students planting the green roof modules on the level 8 balcony.

5.5 Key Challenges: Innovation and Risk

The project did not, however, always run smoothly. One of the key challenges faced in was the risk adverse nature of university operations. Concerns emerged around access to the balcony, user safety, longevity of the roofing membrane, weight loading capacity, and noise controls. A risk management strategy was then developed in collaboration with the project team and university operations team focussing on identifying and assessing risks and developing strategies to manage the risks (Figure 5.8).

The project faced multiple delays as the risks were evaluated and assessed. Strict controls were also put in place including utilisation of raised modules with lockable casters for ease of movement and protections the roof membrane (allowing drains to be easily cleaned), limiting human access near

RISK	DESCRIPTION	LIKELIHOOD (of risk occurring as a result of mitigation strategy)	SEVERITY (of risk occurring as a result of mitigation strategy)	RISK STRATEGY	RISK MITIGATION PLAN
1. Longevity of Water proof membrane	The potential exists for the green roof modules and their layout design to protect the roof membrane from scortching sun and water damage through increase of storm water capture, membrane shading and less trafficable area for pedestrians.	Likely	Minor	Control	Strategic design of the module layout in accordance to sun prone and trafficable.
2. Blockage of Drains due to stormwater flooding	The potential exists to reduce storm water run off through the green roof module design layout, depth and growth medium blend used for optimum absorption and reuse of stormwater	Likely	Minor	Control	Strategic design of module layout (near drainage points but not covering), growth medium depth and blend.
3. Safety Concerns	The potential exists to reduce safety concerns around pedestrian safety through controlled access with a staff member at all times and module positioning against external parapet.	Not Likely	Minor	Control	Strategic design of module layout adjoining the external parapet providing a 3m buffer ensuring inaccessability.
4. Noise Control	The potential exists to reduce noise concerns through limiting pedestrian access to the site through designated opening hours negotiated with staff involved	Not Likely	Minor	Control	Develop a strategy in collaboration with PCPM around potential open hours for module maintenance
5. Movement of Modules due to Fluctuating climatic conditions	The potential exists to reduce the impact of fluctating climatic conditions through securing modules to one another (forming one large contunous module) and using castor breaks ensuring no module movement.	Not Likely	Minor	Control	Design a clip system allowing all modules to connect and therefore create stability during fluctuating climatic conditions. In addition, staff will assess the weather conditions prior to allowing access
6. Structural loading	The potential exists to maximise the existing structural loading capacity through positioning heavier modules over the building columns	Not Likely	Minor	Control	Strategic design of module layout ensuring heavier 'deeper' modules are positioned directly over the building column. Property Services to approve

Figure 5.8   Summary of risk management strategy.

the balcony balustrade through positioning modules against the balustrade to create a buffer zone of 3 m, positioning modules in such a way to ensure a clear line of sight and pathway from one entrance to the other, development of a 'open hours' schedule with building occupants to control noise levels and use of lightweight modules, and growth medium ingredients as well as plant material to ensure the integrity of the structural loading remained.

The project team then used the outcomes of the risk management strategy as an opportunity for transforming the space through integrating the controls as part of the student design competition criteria. Students then transformed these controls into site parameters for design and construction, resulting in a holistic project journey coupled with work integrated learning and an alternative approach to breaking down the barriers for green roof uptake. This was real world learning in action and students learnt valuable lessons about barriers and opportunities to the uptake of their design ideas.

5.6 Conclusion

As a stand alone project, the Matter of Landscape addresses in only a very small way the transition to a LCE. This is a complex challenge, particularly for Universities as key centres for innovation globally. Addressing this challenge, it is argued here, involves more than campus greening initiatives and education about climate change and the reduction of CO_2 emissions. It also requires new ways of thinking about the link between all aspects of University operations. 'New' thinking it is argued here requires spaces for risk taking and innovative thinking, not business as usual. Project-based learning provides one such space. Through a multistage green roof pilot project delivered at RMIT University, methods were developed for harnessing opportunities and addressing barriers to green roof uptake, resulting in institutional change, and further investment and implementation of permanent green roof projects that will provide further research and educational opportunities into the future. Beyond the construction of green roof modules, however, what the project achieved was the development of processes for linking learning and teaching outcomes with university operations and research. It also put in place a learning and teaching space for thinking about and more importantly interrogating taken for granted 'solutions' to address global challenges. It can serve as a model for curriculum innovation as Universities begin to address the demands required and the challenges involved in addressing the consequences of a changing climate.

References

Bartkowicz, K., and Rogers, J. (2015). "Green roofs and urban campus greening: learning about sustainability through doing," in *Integrative Approaches to Sustainable Development at University Level*, eds W. L. Filho, L. Brandli, O. Kuznetsova, and A. M. F. do Paço (Cham: Springer), 495–509.

Brundiers, K., and Wiek, A. (2013). Do we teach what we preach? An international comparative appraisal of problem- and project-based learning courses in sustainability. *Sustainability* 5, 1725–1746.

Claus, K., and Rousseau, S. (2012). Public versus private incentives to invest in green roofs: A cost benefit analysis for Flanders. *Urban For. Urban Green.* 11, 417–425.

Davis, B., and Sumura, D. (2007). 'Complexity science education: reconceptualising the teacher's role. *Learn. Interchange* 38, 53–67.

obson, H. E., and Tomkinson, C. B. (2012). Creating sustainable development change agents through problem-based learning: Designing appropriate student PBL projects. *Int. J. Sustain. High. Educ.* 13, 263–278.

Intergovernmental Panel on Climate Change [IPCC] (2014). *Climate Change 2014: Mitigation of Climate Change.* Available at: https://www.ipcc.ch/report/ar5/wg3/

Leal, W. F., Shiel, C., and Paço, A. (2016). Implementing and operationalising integrative approaches to sustainability in higher education: the role of project-oriented learning. *J. Clean. Prod.* 133, 126–135.

OECD/IEA/NEA/ITF (2015). *Aligning Policies for a Low-carbon Economy.* Paris: OECD Publishing.

RMIT University (2017). *Sustainable Urban Precinct Program.* Available at: https://www.rmit.edu.au/about/our-strategy/values/living-our-values/sustainability/sustainable-urban-precincts-program

Rogers, J. (2013). "Green, brown or grey: green roofs as 'sustainable' infrastructure," in *Sustainable Development and Planning VI*, ed. C. A. Brebbia (Southampton: WIT Press), 323–333.

Shephard, K. (2008). Higher education for sustainability: seeking affective learning outcomes. *Int. J. Sustain. High. Educ.* 9, 87–98

Tourbier, J. T., (2011). *Green Roofs, urban vegetation and urban runoff in The Routledge Handbook of Urban Ecology*, eds. I. Douglas, D. Goode, M. C. Houck, and R. Wang (Oxon: Routledge), 572–582.

United Nations (2015). *Transforming our world: the 2030 Agenda for Sustainable Development.* Available at: https://sustainabledevelopment.un. org/content/documents/21252030%20Agenda%20for%20Sustainable% 20Development%20web.pdf

Wiek, A., Xiong, A., Brundiers, K., and van, D. L. (2014). Integrating problem- and project-based learning into sustainability programs. *Int. J. Sustain. High. Educ.* 15, 431–449.

Wilkinson, S. J., and Reed, R. (2009). Green roof retrofit potential in the central business district. *Property Manage.* 27, 284–301.

World Economic Forum [WEF] (2016). *The Global Risks Report 2016, 11th Edn.* Available at: http://reports.weforum.org/global-risks-2016/

6

Rankings and Sustainability in Portuguese Higher Education Institutions: A Descriptive Analysis

Ana Marta Aleixo[1], Susana Leal[2], Walter Leal Filho[3], Susana Mendes[4] and Ulisses M. Azeiteiro[5]

[1]Universidade Aberta, Portugal
[2]Escola Superior de Gestão e Tecnologia, Instituto Politécnico de Santarém, and Life Quality Research Centre, Portugal
[3]Research and Transfer Centre Applications of Life Sciences, Hamburg University of Applied Sciences, Hamburg, Germany
[4]Escola Superior de Turismo e Tecnologia do Mar, Instituto Politécnico de Leiria, Portugal
[5]Department of Biology and Centre for Environmental and Marine Studies (CESAM), University of Aveiro, Aveiro, Portugal

Abstract

The issue of ranking in Higher Education Institutions (HEIs) has produced a growing body of literature at an international level that is focused on meta-analysis. However, there is still comparably little discussion on the subject of Higher Education (HE) in Portugal. This paper strives to fill this gap. This research was conducted using the Portuguese public HEIs websites and presents a critical review of HEIs rankings in Portugal. These are the types of Higher Education Institution (HEI) rankings that are concerned with the promotion of sustainable development rankings. The links between rankings and institutional commitment, advanced sustainability or the promotion of a positive image are discussed. This work results in the preliminary discussion of a proposal for an alternative ranking for Sustainable Development (SD) in HEIs. An alternative ranking would take the system and subsystem activities of HEIs into consideration, and would constitute a starting point for further discussion when it comes to the development of the ranking for sustainability in HEIs in responding to the issue of holistic and integrated sustainability in Portuguese public HEIs.

6.1 Introduction

Recently, there has been a growing level of reflection on the importance of Higher Education Institutions (HEIs) rankings, allowing distinctions to be made between multiple institutions worldwide. As a result, social responsibility, the impact and quality of scientific research, as well as academic excellence and sustainability have all become key aspects when it comes to distinguishing between HEIs and determining their prestige.

The first rankings date from 1925 in the United States of America, with the next coming about sixty years later in 1981 (Moura & Moura, 2013). However, the Academic Ranking of World Universities, created in 2003, was the first international ranking system (Moura & Moura, 2013). Even though there are now over 33 rankings for HEIs (Shin & Toutkoushian, 2011), few Portuguese HEIs are included in these measurements.

In addition to the Academic Ranking of World Universities, The World University Rankings, QS World University Rankings, Performance Ranking of Scientific Papers for World Universities, Ranking Web of World Universities, CHE- Excellence Ranking, UTD Top 100 Business School Research Ranking are also worldwide rankings with an international perspective.

There has been some criticism about ranking methodologies (e.g., Lozano, 2006; Lukman, Krajnc & Glavic, 2010; Shriberg, 2002) and more needs to be known about the different rankings, their indicators and their reliability. It is necessary to identify these characteristics with a holistic view of the institutions' quality (e.g., Lozano, 2006; Shriberg, 2002). A number of factors have been pointed out that explain the growing acceptance of these rankings by HEIs (e.g., improved image, attracting students, funding, competition, and quality).

International literature reviews on global sustainability rankings in higher education (HE) are scarce and in a Portuguese context, they are non-existent. This paper therefore offers a preliminary discussion about the main SD rankings used in the Portuguese HEIs.

This work also answers the following questions: Which rankings are the Portuguese HEIs indexed to? Are these Institutions indexed in more than one SD ranking?

6.2 Theoretical Context

6.2.1 Genesis and Main Common Indicators of the Current Ranking Used by HEIs

The existing rankings can be categorized in different dimensions, such as internationalization, scientific production and impact/quality, etc. However, it is questionable whether rankings accurately reflect sustainability issues. For example, the Scimago Institutions Rankings[1] is a science evaluation resource that is used to assess universities and research-focused institutions worldwide, using tree different indicators: research, innovation and web visibility indicators. On the other hand, U-Multirank[2], a multidimensional ranking, compares universities with each other on a like-for-like basis in the five broad dimensions of university activity (teaching and learning, research, knowledge transfer, international orientation and regional engagement).

The Center for World University Rankings[3] publishes a global university ranking and measures the quality of education and the training of students as well as the prestige of the faculty members and the quality of their research. This is done without relying on surveys and university data submissions. The Best Global Universities Rankings[4] use 12 indicators that measure academic research performance in addition to global and regional reputation, and were used in two separate ranking indicators (Global research reputation and Regional research reputation).

The CWTS Leiden Ranking[5] is a bibliometric indicator that is based on the Web of Science indexed publications which provides statistics on the scientific impact of universities and on universities' involvement in scientific collaboration. The Academic Ranking of World Universities[6](ARWU) assesses academic performance through the quality of teaching, the quality of academic staff, scientific research, and the performance of academic staff per capita.

The University Ranking by Academic Performance (URAP) ranks world universities based on academic performance determined by the quality and quantity of scholarly publications (citation, article impact and international collaboration).

---

[1] http://www.scimagoir.com/

[2] http://www.umultirank.org/#!/home?trackType=home

[3] http://cwur.org/

[4] http://www.usnews.com/education/best-global-universities/rankings

[5] http://www.leidenranking.com/

[6] http://www.shanghairanking.com/pt/ARWU-Methodology-2015.html

6.2.2 Criticism of Worldwide Rankings in HEIs

The importance of ranking the performance of HEIs has been widely discussed (e.g., Amsler & Bolsmann, 2012; Hazelkorn, 2009; Lukman et al., 2010; Millot, 2015; Shin & Toutkoushian, 2011; Wals, 2014). As stated by Shin and Toutkoushian (2011, p. 1) "with the heightened competition between universities since the 90s and the dramatic growth of the international higher education market, surveys have emerged in many countries as a means of evaluating and ranking universities". According to Shin and Toutkoushian (2011), there are at least 33 rankings that provide objective tools to measure the quality of HEIs and "egalitarianism in higher education".

According to Hazelkorn (2009), the rankings help HEIs to make strategic plans, to respond to competition from other institutions and to strive for excellence. For Lauder, Sari, Suwartha, and Tjahjono (2015), global HEI rankings "are a product of the forces of globalization in higher education", and for Hazelkorn (2009) they help HEIs become "world class". Millot (2015) argues that the international ranking of HEIs is essential nowadays. However, despite their significance and continued use (Lauder et al., 2015), rankings have been harshly criticized by countless researchers. Amsler and Bolsmann (2012) refer to the criticisms of Hazelkorn (2009), namely that rankings benefit some entities over others and increase the gap between them in favor of elite and stratified institutions. For Bookstein, Seidler, Fieder, and Winckler (2010), rankings are not consistent since they vary year after year. Meanwhile, Lauder et al. (2015) noted the disagreement on the quality of education: when evaluating HEIs, some rankings value research, while others focus on reputation and teaching methods. Lauder et al. also note the unfavorable treatment of both smaller institutions and also HEIs that are directed towards humanities and social sciences in which the difference in research levels raises many difficulties.

Lukman et al. (2010) note that rankings are based on subjective qualitative data provided by the HEIs and often place too much emphasis on research. In this context, Lukman et al. (2010) introduced a new indicator which values research, education and the environment.

Wals (2014) states that HEIs engaging in sustainability are generally institutions that focus on education rather than on research, whereas HEIs with better research pay less attention to Education for Sustainable Development and sustainability.

In this context, SNAHE states that "Ranking is an attempt to measure the quality of higher education or research, but it differs from many other forms of quality assessment because it is relative – there are no absolute criteria or norms for what may be regarded as "minimum quality"(SNAHE, 2009, p. 11).

According to PRME (2015), rankings have limitations and can lack transparency and rigor even though many schools (business schools, for example) depend on them in some way. Isaksson and Johnson (2013) note that HEI rankings can play a role when students and their parents are choosing a university. Rankings foster public awareness and encourage transparency in Higher Education (Millot, 2015).

6.2.3 The Impact and Influence of Rankings

The benefits associated with joining rankings include the following: (i) national and international reputation, (ii) attracting a larger number of students, (iii) obtaining external financing, (iv) training better professionals, (v) evaluating performance, (vi) inter-institutional competition, (vii) enabling strategic planning, (viii) quality control, (ix) reorganizing and restructuring institutions and (x) influencing curriculum and training priorities (Hazelkorn, 2009). Regarding advertising and the favorable image which may result from joining these rankings, Cebrian, Grace, and Humphris (2015, p. 79) quote Lukman et al. (2010) "University world rankings put in place benchmarking systems that influence decision-making process". On the other hand, the disadvantages that arise when an institution joins rankings are related to (i) potential negative effects on their reputation and (ii) the financial risks associated with a dependence on foreign students.

Moura and Moura (2013) also identify the forces that might determine HEIs' choice of rankings, such as the prestige they grant to their staff, the access to funds for research, the increased demand for R&D (particularly from the industry), and the recruitment of the best researchers, teachers and national and international students.

Ranking systems will help to determine the choices made by different groups (e.g., students, teachers and external entities) in cases such as the selection made by students and families, teachers who want to further their training (e.g. Doctoral and post-doctoral), as well as external entities from which HEIs can obtain financing. On the other hand, Moura and Moura (2013) refer to the rankings' tendency to homogenize HEIs, which in turn respond by developing their activities and targets according to ranking levels.

Institutions therefore assume behaviors within the system that might not necessarily respond to the needs and specific interests of students and staff, but rather might help their performance in global rankings.

HEIs have increasingly joined these rankings due to the influence of factors like performance, quality and excellence. Different researchers (e.g., Lukman et al., 2010; Saisana, d'Hombres, & Saltelli, 2011) note that the Academic Ranking of World Universities (ARWU) and the Times Higher Education (THE) attract the attention of HEIs worldwide.

What drives HEIs to join these rankings? Although there is no easy answer to this question, issues such as positive publicity, institutional commitment or development, and sustainability may provide some answers.

The SNAHE (2009) draws attention to the many arguments for HEIs adopting rankings. Above all, they provide free publicity by informing students and external entities about quality standards, education and higher education research. According to Bernardino and Marques (2010, p. 31), the rankings could "give bolster the reputation of a university and provide free publicity to institutions". However, these authors criticize the illusory image and quality which encourages unreasonable behaviors from both students and institutions. Bookstein et al. (2010) also demonstrate unacceptably high fluctuations in yearly rankings.

For Shin and Toutkoushian (2011), rankings help consumers see the value of investment, hold institutions accountable for results, provide students and parents with a comparison between institutions, showcase high quality institutions for teaching and research, and contribute to the development of society. The Berlin Principles of Higher Education Institutions[7] set out the implications for the future of ranking and transparency, relevancy, and the validity of comparative data (Shin & Toutkoushian, 2011).

6.2.4 Historical Background of Ranking in Portuguese HEIs

In Portugal, only a few HEIs are included in rankings (namely universities which use them to showcase their quality), and some are now only just beginning to do so. Still, the number is growing and their commitment to sustainability focused rankings is becoming more visible.

Bernardino and Marques (2010, p. 31) state that "in several international reviews and evaluations of the Portuguese higher education it was pointed out

---

[7]http://www.che.de/downloads/Berlin_Principles_IREG_534.pdf (11.01.2015).

that the Portuguese institutions do not provide the minimum information to stakeholders, and that they were not competitive enough". And according to Shi and Lai (2013), the ranking systems could contribute to HEI managers adopting SD.

6.2.5 The Emergence of Sustainability Rankings

For Lukman et al. (2010, p. 621) "the implication of environmental issues has received little or no attention at all, although many universities are monitoring their environmental footprint". In this context, Lauder et al. (2015) mention that the global sustainability ranking in higher education is still poorly developed.

According to Wals (2014), not enough attention is given to sustainability by HEIs despite the importance of these rankings. They mostly focus on research, internationalization, student evaluation or external research funding. Wals claims that only The Green League Table[8] uses a range of sustainability indicators (Wals, 2014).

PRME (2015, p. 8) suggests that despite the existence of a large number of sustainability rankings, few cover the full scope of SD and often refer only to environmental issues on campus. Further, he mentions that "better integration of sustainability into current well-respected rankings would be an incentive for schools to change".

The UI Green Metric Ranking of World Universities (UI GREEN-METRIC 2015) is a global ranking for HEIs developed by the University of Indonesia. Its development was based on ideas taken from the International Conference on World University Rankings and quality of Education (Lauder et al., 2015). It covers the following items: setting and infrastructure, energy and climate change, waste, water, transportation, and education. The latter was introduced a year after the other items (Lauder et al., 2015).

As mentioned by Cebrian et al. (2015) and Lauder et al. (2015), ranking HEIs in the field of sustainability could foster a holistic implementation of SD in HEIs, and the GreenMetric is an example of this. According to both Cebrian et al. (2015) and Lauder et al. (2015), the GreenMetric could be a tool to integrate sustainability assessment and reporting methods in HEIs. This is also suggested by Grindsted (2011), who states that the UI ranking is the first to reflect HEI behavior when it comes to sustainability.

---

[8]http://www.thecompleteuniversityguide.co.uk/league-tables (11.01.2016).

On the other hand, Lozano (2006) claims that most international instrument frameworks for measuring sustainability in HEIs do not include education or research, and their evaluation is called into question. Lozano (2006) states that some tools do not give a complete picture or consider social dimensions. Rather, they either address environmental aspects, or they attribute different weights to each dimension of SD. (e.g., Global Reporting Initiative, ISO 14000, Ecological footprint . . .).

Therefore, Lozano (2006) believes that although the ISO 14000 offers a three-part certification (environmental, economic and social) and international recognition, it puts less emphasis on social and economic dimensions than it does for the environment. He notes that the Graphical Assessment of Sustainability in Universities (GASU) answers two questions: i) Where do institutions stand regarding sustainability and what efforts have they established? ii) How do they communicate these efforts to stakeholders and their progress in the various dimensions of sustainability? Subsequently, for Lozano (2006), an effective tool should be easy to measure, allowing for the comparison of different HEIs with the same weights and measures, while Shriberg (2002) claims it should encompass an ideal assessment area, open to multiple stakeholders.

Lozano (2011) underlined the importance of the educational dimension to sustainability reporting and it was subsequently added to the economic, environmental and social dimensions.

According to Disterheft et al. (2016, p. 173) "concerns were expressed that sustainability assessment practices in HE run the risk of catering more towards market demands than to societal needs and transformative change, in particular when they focus on quantitative oriented ranking systems".

Considering that less is known about the rankings adopted by Portuguese HEIs, we intend to analyze the main rankings adopted, as well as to determine if Portuguese HEIs participate in more than one ranking.

6.3 Methods

6.3.1 Content Analysis

A content analysis of all Portuguese public HEI websites was carried out in order to measure the ranking information disclosed by Portuguese HEIs. This involved categorizing the disclosed information. A simpler analysis was also conducted in order to detect the presence or absence of rankings adopted by Portuguese HEIs. This method has been successfully advocated in the literature reviewed (e.g., Amaral, Martins, & Gouveia, 2015; Brondani, Brandli, Frandoloso, & Vieira, 2014; Katiliūtė, Daunorienė, & Katkutė, 2014; Lozano, 2010). The data collection was performed through the use of publicly accessible data, available on the main HEI websites. Data collection took place from August 2015, and each website from the 34 HEIs was manually reviewed.

6.3.2 Sample

This study reviewed the 34 public Portuguese HEIs[9] (20 of which are poly-technics and 14 are universities) institutional websites (the main institutional website of each university or polytechnic).

6.3.3 Procedures

In the analysis of the HEI websites, the different variables are coded as fol-lows: firstly, identified rankings were coded as 1; unidentified rankings were coded as 0. Data that was collected via the existing ranking databases was also collected. Secondly, both documents (in excel format) were compared. The first, complete with ranking data disclosed on the official HEI webpages, and the other, presenting data that was collected via the existing ranking databases. Lastly, the initial code was used again: the identified rankings were coded as 1, and the unidentified ones were coded as 0.

6.4 Results

Most public Portuguese HEIs are not listed in the rankings (n = 13; 38,2%; Figure 6.1), six HEIs are listed in one ranking (n = 6; 17,7%), while another six are listed in two rankings (n = 6; 17,7%). The remaining number of rankings adopted by each public Portuguese HEIs are: three rankings are

---

[9]By the end of 2014, the Portuguese network of public HEIs was made up of fourteen Universities, twenty Polytechnic Institutes, and eight Military and Police Higher Education schools. In this study, we only considered the Universities and the Polytechnics. It should be noted, however, that some polytechnic schools integrated into the university system (14 poly-technic schools were integrated into six universities), which were included in the university domain.

Figure 6.1   Histogram of the number of rankings adopted by each public Portuguese HEI (n = 34).

adopted by two HEIs, four rankings are adopted by two HEIs, six and seven rankings are adopted by one HEI each, and eight rankings are adopted by three HEIs.

The rankings used to disclose information by public Portuguese HEIs are listed in Figure 6.2. The rankings more frequently adopted are: Scimago (n = 19 HEIs; 55,9%), University Ranking of Academic Performance (n = 13; 38,2%), U-Multirank (n = 10; 29,4%). Green Metric, an SD ranking, is used only by two HEIs (one university and one polytechnic).

Although there may be more rankings in HEIs than those presented herein, the purpose of this study was to understand how HEIs publicize the adoption of rankings in their websites.

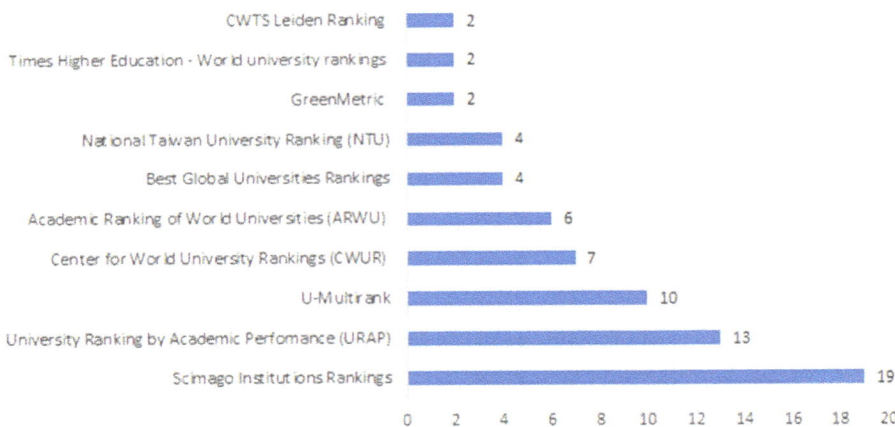

Figure 6.2   Rankings adopted by each public Portuguese HEIs (n = 34).

6.5 Discussion

A comprehensive analysis of indicators/dimensions considered in the diffe-rent rankings revealed that while there are many different indicators, these refer to similar concepts, such as *International Collaboration in the Best Global Universities Rankings* and International Outlook in the *The Times Higher Education World Universities Ranking.* Moreover, others are incom-parable, for example *Total Citations in the Best Global Universities Ranking* vs. *Citations in the Center for World University Ranking.*

Some of the indicators (e.g., Research, International Collaboration, Knowledge Transfer and Publications) correspond to the institutional, politi-cal or educational dimensions structured by Aleixo et al. (2016) as proposed by several authors such as Lozano (2010, 2011) and Leal Filho, Manolas & Pace (2015). However, environmental, economic and social dimensions are not identified. According Aleixo et al. (2016) an institutional dimen-sion includes: (i) The declarations and actions that explain how the HEI views values, strategy, transparency in governance and ethical compromises; (ii) The declarations, explanations, and links with the national and interna-tional criteria on aspects of sustainable development, and (iii) The practices in education, research, university operations (e.g. certifications), the external community and assessment and reporting.

Accordingly, some of the indicators used in the rankings may have some resemblance to practices in education, research and assessment. Thereby, existing rankings used by Portuguese HEIs, with the exception of Green-Metric, do not integrate sustainability issues. Other indicators are privileged, namely those regarding the numbers of faculties and students or other research indicators.

In view of the above, and because the two rankings might not have equality measures, we recommend the GreenMetric with the addition of a further four dimensions of sustainability, as defended by Aleixo et al. (2016).

Despite the large variety of rankings, none can fully portray the develop-ment of sustainability in HEIs. Therefore, in the face of monolithic rankings, it is proposed that future rankings provide a holistic and integrated ranking of the four dimensions of SD (environmental, social, economic and institutional dimensions) (Aleixo et al., 2016). In this way it is possible to respond to the objective of sustainable HEIs. The proposed ranking should also include all activities of the HEIs, including research and education. Most rankings do not include all of these activities systemically, in a proportionate manner of analysis/evaluation, namely: (i) education, (ii) research, (iii) campus opera-tions, (iv) community outreach, (v) assessment and reporting.

The GreenMetric answers notable concerns regarding the environment (environmental issues on campus) as well as institutional dimensions (education, research and events for sustainability). However, the sustainability of a society and the role that higher education institutions plays in relation to this purpose should include other concerns, particularly when it comes to those of a social and economic nature. A new proposal for a ranking should fully integrate all four dimensions and all activities in HEIs, as for example, was proposed by Lozano et al (2015) in their survey: (i) SD education (e.g., courses for SD), (ii) SD research (e.g., research units for SD), (iii) outreach and community (e.g., provides services to the community, exchange programmes in field of SD), (iv) campus operations (e.g., sustainable buildings, separation and recycling, energy renewable), (v) institutional commitment and framework (e.g., SD working group, mission for SD), and (vi) assessment and reporting (e.g., sustainability reports and sustainability rankings).

6.6 Conclusions

In this paper, we aimed to describe the adoption of rankings by public Portuguese HEIs. Portuguese Higher Education Institutions use these rankings to showcase their quality. The Times Higher Education Supplement (THES) and the Shanghai Jiao Tong University (SJTU) (Lukman et al., 2010, p. 621) are the most frequently used as they are "more influential worldwide" and follow different methodologies: the former is based on peer review and the latter focuses on research outputs (Bernardino & Marques, 2010). However, in Portugal, the most used rankings are the Scimago and the University Ranking of Academic Performance.

With this study, we identified a heterogeneous approach for the rankings used. Although few HEIs adopted six to eight rankings, most of the HEIs are not listed in any ranking at all. The use of SD rankings is even more scarce.

Lauder et al. (2015) presented a table featuring the users and stakeholders for academic and sustainability rankings. According to this analysis, factors such as infrastructure, energy, climate change and water and transportation waste are fundamental in the development of sustainability rankings, as opposed to factors like research, teaching, reputation and internationalization.

Therefore, a comprehensive ranking encompassing all SD dimensions, in accordance with the Aleixo et al. (2016) proposal, could constitute a unifying proposal for HEIs. Such a ranking would therefore include, for example, the following dimensions: SD research, SD education and institution internationalization. However, the assessment items in each of the dimensions should be monitored in order to identify the possible need for new topics in the aspects of sustainable development. This new ranking would provide HEIs new possibilities in terms of sustainability and would constitute an unbiased evaluation tool for Portuguese HEIs.

Although the ranking cannot be used to predict university sustainability performance (Isaksson & Johnson, 2013), the HEIs should choose rankings that include a sustainability assessment in order to demonstrate their purpose, mission, quality and so on. HEIs should see the ranking as a way to demonstrate to citizens the quality of their education and should not simply treat rankings as a benchmark tool. In order address criticisms by Lauder et al. (2015) about the lack of reviews of global sustainability rankings in HEIs, a specialized team is required to provide a better designed sustainability ranking that can make a genuine contribution to sustainability and to society as a whole.

References

Aleixo, A. M., Azeiteiro, U. M., and Leal, S. (2016). "Toward sustainability through higher education: sustainable development incorporation into portuguese higher education institutions," in *Challenges in Higher Education for Sustainability*, eds J. P. Davim and W. Leal Filho (London: Springer), 159–187.

Amaral, L. P., Martins, N., and Gouveia, J. B. (2015). Quest for a sustainable university: a review. *Int. J. Sustain. High. Educ.* 16, 155–172. doi: 10.1108/IJSHE-02-2013-0017

Amsler, S. S., and Bolsmann, C. (2012). University ranking as social exclusion. *Br. J. Sociol. Educ.* 33, 283–301. doi: 10.1080/01425692.2011.649835

Bernardino, P., and Marques, R. C. (2010). Academic rankings: an approach to rank portuguese universities. *Ensaio* 18, 29–48.

Bookstein, F. L., Seidler, H., Fieder, M., and Winckler, G. (2010). Too much noise in the Times Higher Education rankings. *Scientometrics* 85, 295–299. doi: 10.1007/s11192-010-0189-5

Brondani, S. C., Brandli, L. L., Frandoloso, M. A. L., and Vieira, S. (2014). Panorama da sustentabilidade ambiental nas melhores universidades da américa latina. *Rev. Aidis Ingeniería Ciên. Ambientales* 7, 1–10.

Cebrian, G., Grace, M., and Humphris, D. (2015). Academic staff engagement in education for sustainable development. *J. Clean. Prod.* 106, 79–86.

Disterheft, A., Caeiro, S., Leal Filho, W., and Azeiteiro, U. M. (2016). The INDICARE-model – measuring and caring about participation inhigher education's sustainability assessment. *Ecol. Indic.* 63, 172–186. doi: 10.1016/j.ecolind.2015.11.057

Grindsted, T. S. (2011). Sustainable universities: from declarations on sustainability in higher education to national law. *Environ. Econ.* 2, 29–36.

Hazelkorn, E. (2009). Rankings and the battle for world-class excellence: institutional strategies and policy choices. *High. Educ. Manage. Policy* 21, 1–22. Isaksson, R., and Johnson, M. (2013). A preliminary model for assessing university sustainability from the student perspective. *Sustainability* 5, 3690–3701. doi: 10.3390/su5093690

Katiliūtė, E., Daunorienė, A., and Katkutė, J. (2014). Communicating the sustainability issues in higher education institutions World Wide Webs. *Proced. Soc. Behav. Sci.* 156, 106–110. doi: 10.1016/j.sbspro.2014.11.129

Lauder, A., Sari, R. F., Suwartha, N., and Tjahjono, G. (2015). Critical review of a global campus sustainability ranking: GreenMetric. *J. Clean. Prod.* 108, 852–863. doi: 10.1016/j.jclepro.2015.02.080

Leal Filho, W., Manolas, E., and Pace, P. (2015). The future we want: key issues on sustainable development in higher education after Rio and the UN decade of education for sustainable development. *Int. J. Sustain. High. Educ.* 16, 11–129. doi: 10.1108/IJSHE-03-2014-0036

Lozano, R. (2006). Incorporation and institutionalization of SD into universities: breaking through barriers to change. *J. Clean. Prod.* 14, 787–796.

Lozano, R. (2010). Diffusion of sustainable development in universities curricula: an empirical example from Cardiff University. *J. Clean Prod.* 18, 637–644. doi: 10.1016/j.jclepro.2009.07.005

Lozano, R. (2011). The state of sustainability reporting in universities. *Int. J. Sustain. High. Educ.* 12, 67–78. doi: 10.1108/1467637111098311

Lozano, R., Ceulemans, K., Alonso-Almeida, M., Huisingh, D., Lozano, F. J., Waas, T., Lambrechts, W., Lukman, R., and Hug, J., (2015). A review of commitment and implementation of sustainable development in higher education: results from a worldwide survey. *J. Clean. Prod.* 108, 1–18. doi: 10.1016/j.jclepro.2014.09.048

Lukman, R., Krajnc, D., and Glavic, P. (2010). University ranking using research, educational and environmental indicators. *J. Clean. Prod.* 18, 619–628. doi: 10.1016/j.jclepro.2009.09.015

Millot, B. (2015). International rankings: Universities vs. higher education systems. *Int. J. Educ. Dev.* 40, 156–165. doi: 10.1016/j.ijedudev. 2014.10.004

Moura, B. A., and Moura, L. B. A. (2013). Ranqueamento de universidades: reflexões acerca da construção de reconhecimento institucional. *Acta Sci. Educ.* 35, 213–222. doi: 10.4025/actascieduc.v35i2.20400

PRME (2015). State of Sustainability in Management Education. *Global Compact LEAD and PRME.*

Saisana, M., d'Hombres, B., and Saltelli, A. (2011). Rickety numbers: Volatility of universities rankings and policy implications. *Res. Policy* 40, 165–177. doi: 10.1016/j.respol.2010.09.003

Shi, H., and Lai, E. (2013). An alternative university sustainability rating framework with a structured criteria tree. *J. Clean. Prod.* 61, 59–69. doi: 10.1016/j.jclepro.2013.09.006

Shin, J., and Toutkoushian, R. (2011). "The past, present, and future of university ranking," in *University Ranking: Theoretical Basis, Methodology and Impacts on Global Higher Education*, eds J. Shin and U. Teichler (Dordrecht: Springer).

Shriberg, M. (2002). Institutional assessment tools for sustainability in higher education. *Int. J. Sustain. High. Educ.* 3, 254–270. doi: 10.1108/14676370210434714

SNAHE (2009). *Ranking of Universities and Higher Education Institutions for Student Information Purposes?*

UI GREENMETRIC (2015). *UI Greenmetric World University Ranking – Guideline.* Available at: http://greenmetric.ui.ac.id/wp-content/uploads/2015/07/UI_Greenmetric_Guideline_2015.pdf [accessed August 12, 2015].

Wals, A. E. J. (2014). Sustainability in higher education in the context of the UN DESD: a review of learning and institutionalization processes. *J. Clean. Prod.* 62, 8–15. doi: 10.1016/j.jclepro.2013.06.007

7

Pathways to a Low Carbon Economy: The Evolving Role of the University of the Philippines

**Mark Anthony M. Gamboa, Kristine F. Aspiras
and Annlouise Genevieve M. Castro**

Faculty Room 4, School of Urban and Regional Planning,
University of the Philippines, Diliman, E. Jacinto St.,
UP Diliman Campus, Quezon City, Philippines

Abstract

This article assesses the evolving role of the University of the Philippines System, which is the only national university in the country, in the transition of higher education institutions to a low-carbon economy. As national university, various initiatives have been put in place as it is duty-bound to contribute to sustainable development. Foremost is the University's Green UP program that aims to promulgate policies across the system, including green building standards, energy audits, replacement of equipment to improve energy efficiency, water and solid waste management, monitoring of building maintenance standards, environmental management, and other related strategic initiatives in relation with climate change, disaster risk reduction, and sustainability.

7.1 Introduction

Institutions of higher education vary greatly, from community colleges, to small private and public liberal arts colleges, to large private and public research universities (Sungu-Eryilmaz, 2009). Perry and Wiewel (2005) emphasize the "undeniable importance of the roles universities play in the growth and development of the city". The university is considered as a center of culture, aesthetic direction, and the moral forces shaping the "civilized" society. It contributes in important ways to economic health and physical landscape of cities, serving as all but permanent fixtures of the urban economy and built environment. While once functioning mainly as enclaves of intellectual pursuit, colleges, and universities today play a much broader role in the economic, social, and physical development of their host cities and neighborhoods. They have become key institutions, often termed anchor institutions, in their communities through their economic impacts on employment, spending, and work-force development, as well as through their ability to attract new businesses and highly skilled individuals and to revitalize adjacent neighborhoods. Among their many economic impacts, the most important ones are enhancing the industry and technology base, employing large numbers of people, and generating revenue for local governments through university expenditures on salaries, goods, and services.

In the Philippines, a total of 560 public higher education institutions (including satellite campus) and 1,710 private higher education institutions are reported for academic year 2016–2017. The administrative regions with the highest presence of higher education institutions (HEIs) include the National Capital Region, CALABARZON, and Central Luzon. These are the regions that belong to the Greater Capital Region (Tables 7.1 and 7.2).

For public HEIs, the Roadmap for Public Higher Education Reform envisions state universities and colleges (SUCs) to be contributing significantly to "the urgent tasks of alleviating poverty, hastening the pace of innovations, creating new knowledge and functional skills, and increasing the

Table 7.1   Distribution of higher education institutions in the Philippines by institution type

Indicator	2014–15	2015–16	2016–17
Total HEIs (excluding state universities and colleges (SUCs) Satellite campuses)	1,935	1,934	1,943
Total HEIs (including SUCs Satellite campuses)	2,388	2,388	2,396
Public	227	228	233
SUCs	112	112	112
SUCs Satellite Campuses	453	454	453
Local Colleges and Universities (LUCs)	101	102	107
Others (include OGS, CSI, Special HEI)	14	14	14
Private	1,708	1,706	1,710
Sectarian	360	359	351
Non-Sectarian	1,348	1,347	1,359
Autonomous/Deregulated Private HEIs	64	75	75

Source: Commission on Higher Education (2017).

Table 7.2 Distribution of higher education in the Philippines by region: AY 2016–2017

	Region	Count
1	Ilocos Region	92
2	Cagayan Valley	53
3	Central Luzon	201
4	CALABARZON	283
5	Bicol Region	141
6	Western Visayas	64
7	Central Visayas	117
8	Eastern Visayas	65
9	Zamboanga Peninsula	61
10	Northern Mindanao	76
11	Davao Region	86
12	Soccsksargen	101
13	National Capital Region	345
14	Cordillera Administrative Region	43
15	Autonomous Region of Muslim Mindanao	65
16	Caraga	43
17	MIMAROPA	49
18	Negros Island Region	58
	Total	1,943

Data exclude satellite campus.

Source: Commission on Higher Education (2017).

productivity of the workplace and the dynamism of communities" through the triad functions of instruction, research, and extension (Commission on Higher Education, 2012).

The University of the Philippines System comprises eight constituent universities spread across the country. Declared by Republic Act No. 9500 (otherwise known as the UP Charter of 2008), it is the only public national university, placing it in a unique position as a catalyst towards transforming HEIs to a low-carbon economy (LCE) particularly in building resilience through synergies in education.

7.2 UP as a Green University: Innate Characteristics and Early Traditions

The University has always upheld the integrity of the natural environment. Over the years, it has demonstrated this respect in a number of its policies and principles from its land use plans, campus design and layout, and even the lifestyle of its communities.

The University's primary policy document pertinent to the promotion of low carbon campuses is the University of the Philippines Master Development Plan: Development Principles and Design Guidelines (UPMDP) prepared by the Office of Design and Planning Initiatives-Office of the Vice President for Development in 2014 and approved by the UP Board of Regents at its 1300th meeting on 28 August, 2014. The policy document was formulated with the active participation of the University's constituents and key stakeholders through various consultations and workshops (ODPI-OVPD, 2014). The UPMDP calls for the formulation of a Campus Master Plan (UPCMP) that will serve as the framework for all developments in the constituent universities and campuses of the University of the Philippines System. The UPCMP aims "to stimulate, govern, and control development, designate land-use zones and indicate road and pedestrian networks so as to guide the development and ensure the creation of the University of the Philippines which is socially responsible, functional, innovative, and visionary" (ODPI-OVPD, 2014). This statement embodies the stated vision in the U.P. Strategic Plan 2011–2017, i.e., *"A great university, taking a leadership role in the development of a globally competitive Philippines"* (ODPI-OVPD, 2014).

The UPCMP aims to ensure that the university will maintain low-carbon campuses with environmentally sustainable transport systems with efficient levels of services, anchored on ensuring inclusive mobility and the promotion of non-motorized transport in the various campuses. In particular, the UPCMD identifies nineteen (19) master development planning principles with the following principles as most relevant to the promotion of a low-carbon campuses:

- *Green UP: environmentally sustainable and risk-sensitive design.* Promote environmentally sustainable and green architecture design, aimed at reducing the negative impacts of the construction of buildings on the natural environment and at promoting the comfort, safety and wellbeing of its users.
- *Pedestrian and bicycle friendly community.* Encourage walking and biking as a pleasurable means of transportation by providing the safety devices and infrastructure for these environmentally friendly activities, such as bicycle paths and bicycle parking, and pedestrian walkways, footpaths and sidewalks. This is consistent with the promotion of low-carbon campuses through the adoption of environmentally sustainable and non-motorized transportation systems.

- *Operational efficiency.* Formulate policies and programs that mandate or provide incentives to constituent units that implement energy efficiency programs in their utility operations, thereby reducing total energy expenditures and improving energy efficiency awareness system-wide. Incorporate innovative energy efficient technologies in the overall building and utilities design and planning to achieve improvements in utility generation, transmission, and distribution.
- *Promotion of urban agriculture.* Contribute to food security and food safety, through bio-intensive and energy-saving food production methods in non-productive land areas, by increasing the amount of food available, such as fresh vegetables, fruits, and meat products, to campus constituents and other people living in the vicinity (as cited in ODPI-OVPD, 2014).

The UPMDP adapted the American Institute of Architects' principles for livable communities to complement the planning and design of the University's academic campuses (ODPI-OVPD, 2014). The document promotes the principle of building on a human scale wherein the campus layout is "compact, pedestrian friendly and designed to match human scale" (ODPI-OVPD, 2014). It also promotes the provision of transportation options to the UP community whether it be by walking, biking or motorized transportation through the formulation of a responsive transport and traffic management plans at various campuses to ensure low-carbon environmentally sustainable transportation (ODPI-OVPD, 2014). In terms of the design of buildings and sites, the Guidelines calls for buildings and sites that are functional, structurally safe, pleasing to the eyes, and sustainable (ODPI-OVPD, 2014).

The detailed land use plan takes into consideration the highest and best use (optimum) of the land with activities and its locations delineated. As stated in the UPMDP, the elements (i.e., allowable land uses) that have to be incorporated in the layout and character of the campus include the following:

- *Campus Core*: the historic and unifying center of the campus – with appropriately maintained pioneer buildings, heritage trees, and other campus elements – that shall become the inspiration of all future developments on campus;
- *Programmed Open Spaces*: large tracts of campus green spaces, integrated with softscape and hardscape in a designed exterior environment, which allows for a variety of human activities, both passive and active;
- *Protected Natural Open Spaces*: designated zones of natural or man-made forests, waterways, wetlands and geo-hazard areas which shall

Figure 7.1 The Facade of Quezon Hall: a pioneer building in UP Diliman (*Photo Credit*: M.A.M. Gamboa, 2015).

remain untouched and protected, in accordance with law and University rules and regulations;

- *Agricultural Zones*: include expanses or urban land and wetlands that are preserved and protected for agricultural production and educational purposes.

Figure 7.2 The University Amphitheater where commencement exercises, parades, and other activities are usually held (*Photo Credit*: M.A.M. Gamboa, 2015).

Figure 7.3 The UP Lagoon and its vegetation (*Photo Credit*: M.A.M. Gamboa, 2015).

Furthermore, the University continues to encourage walking and biking as a pleasurable means of transportation by providing the safety devices and infrastructure for these environmentally friendly activities, such as bicycle paths and bicycle parking, and pedestrian walkways, footpaths, and sidewalks (Figures 7.3 and 7.4). It is also implementing a carless scheme along its

Figure 7.4 UP Diliman's Academic Oval with designated pedestrian, non-motorized, and motorized vehicle lanes (*Photo Credit*: M.A.M. Gamboa, 2015).

Figure 7.5 UP Diliman's Automated Guideway Transit System a prototype of a zero-carbon emission train (*Photo Credit*: M.A.M. Gamboa, 2015).

academic highway, which is the Academic Oval, every Sundays in its Diliman campus. In terms of moving towards green transportation means, the University has embarked on the construction of the UP-Automated Guideway Transit System (Figure 7.5), which is a joint project with the Department of Science and Technology and the use of electronic vehicles, more particularly electronic tricycles, as part of its research initiatives on environment friendly and sustainable modes of transportation within its campuses.

7.3 Green UP Program

The Green UP is a program initiated by the University of the Philippines to bring together the efforts of the UP Community in creating UP Campuses that are environment friendly and resilient to climate change impacts. The Green UP program, as a collective expression of the university's commitment to sustainable development, highlights on researches, technology, and best practices of the various campuses.

In support of the Green UP program, each of the Constituent Universities (CUs) has initiated their own campus-wide policies and translating them into activities that lead towards a healthier and less polluted environment. All CUs have now incorporated aspects of sustainable development into their respective master plans. All campuses maintain their own campus-wide policies that relate to Green UP, which they have translated into programs, projects, and activities in the areas of research, operations, instruction, extension, policy and infrastructure, within their campuses, as reflected in Table 7.3. These initiatives reflect the role of the University not just in building knowledge and educating future low-carbon leaders in the country and in the region but also in ensuring that the campuses of the University themselves serve as model of low carbon development by ensuring that the University's operations adhere to the principles and practices of a low carbon economy (LCE).

As part of the Green UP initiative, the University has also entered into agreements with private institutions for the installation of solar photovoltaic systems on the roofs of a number of its buildings. Moreover, the design of new buildings, such as the UP Integrated School, and its new campuses, such as the UP Clark Green City and the UP Tacloban-Sta Elena Campus, now integrate the principles of green building and resiliency.

7.4 The University and Its Contribution to Its Wider Region

The University of the Philippines is also a public service university both by virtue of the law (RA 9500 or the UP Charter) and its century-old tradition of providing service to the Filipino nation. Thus, apart from the individual efforts and targets of UP Constituents to contribute to the Green UP Program, numerous programs and projects have been undertaken in partnership with other government agencies and organizations. These include the following most current endeavors:

7.4.1 Project SARAI

An undertaking of the University of Philippines Los Baños (UPLB) and the Department of Science and Technology – Philippine Council for Agriculture, Aquatic and Natural Resources Research and Development (DOST-PCAARRD), Project SARAI or the Smarter Approaches to Reinvigorate Agriculture as an Industry is intended to come up with a national crop forecasting and monitoring service for six priority crops, namely: rice, corn, banana, coconut, coffee, and cacao. Part of the project's objectives is to empower farmers with smarter options through state-of-the-art technologies and near real-time crop advisories. A list of tools and applications has likewise already been developed such as the Suitability Map Application,

Table 7.3 Summary of initiatives of UP system's constituent universities towards a low-carbon campus

Research	Operations	Extension	Instruction	Policies	Infrastructure
UP Baguio: • Rainwater Harvesting • PhiLidar 1; innovative methods of gathering information for disaster mitigation • New Materials and Devices for Energy Harvesting and Storage	UP Baguio: • Carless Oval (OVCCA, 2010) • Car Sticker System (OVCCA, 2010) • Introduction of E-Trike or E-Kot • Paperless Transactions • LED Lights UP Manila: • Waste Segregation UP Los Baños: • UPLB Energy Audit • UPLB Solid Waste Management Program • Energy Conservation Program (austerity in the use of aircon)	UP Diliman: • Taskforce on Solid Waste Management (OVCCA, 2000) • Activities from Academic Units UP Visayas: • Site development and reforestation project • Tree planting/greening program with the DENR • Landscaping • Community-based Bamboo Enterprise • Diploma in Urban and Regional Planning from the College of Management	UP Baguio: • Curriculum development including the establishment of MS in Conservation and Restoration Ecology (MS CARE); Proposal for MS/PhD Sustainable Science Program	UP Baguio: • Green Governance UP Diliman: • Green Building Policies in new infrastructure • Prohibition on the use of Styrofoam/plastic UP Cebu: • Local policy: required pollution control officer (no officer, no garbage collection) • Compliance with green space UP Mindanao: • Promoting use of bicycles within the campus	UP Baguio: • Mini-agroforestry farm and Narra plantation around UP BREHA • Medical Herb Garden inside campus • Arboretum of native trees at Sabkil, the 4-hectare UP Baguio at Itogon, Beguet • Museo Kordilerya (Cordillera Ethnographic Museum), Kolehiyo ng Agham Research Lab, CAC Theatre, HKP Gymnasium, BJL Renovation, Balay International UP Diliman: • Revitalization of UP Arboretum

Table 7.3 Continued

UP Diliman: • Source of Solutions Grants (SOS) UP Cebu: • LIDAR (partnership with DOST and West Visayas LGUs) UP Visayas: • Consortium on maritime regimes and ecosystems and marine life, looking at coastal resource management UP Mindanao: • Exploratory research work on: • Bamboo materials properties and construction	• UPLB Organic Programs (replacement/minimal use of inorganic fertilizers) UP Cebu: • Hazard Mapping UP Visayas: • Power conservation and use of renewal energy source • Landfill for solid waste management • Facultative pot for waste water limited treatment UP Mindanao: • Implementation of waste management segregation in the university and at the barangay level	UP System: • Tree planting activities under the Environmental Corporate Responsibility	• Green Building Project UP Los Baños: • Alternative Energy • Alternative Feedstock UP Open University: • Materials Recovery Facility • Solar Street Lamps • Retrofitting buildings with T5 fluorescent tubes • Green Building Community Hub UP Cebu: • MOA with Ayala (Asia Town IT Park) UP Mindanao: • Infrastructure/Construction • Incorporate natural lighting and ventilation in buildings • Retrofitting of existing buildings to include a

(Continued)

Table 7.3 Continued

Research	Operations	Extension	Instruction	Policies	Infrastructure
• Organic waste • Paper-crete/eco-masonry units • Composite construction materials using agricultural waste	• Observe proper procedures on tree preservation and inventory • 9:00 AM to 4:00 PM use of aircon units				rainwater collection system • Procurement and use of LED lights in new buildings • Procurement of inverter air conditioning units • Incorporate sola tubes in New Training Gym UP System: • Design and construct bicycle parking, showers, and lockers for all new buildings to be constructed within the campuses • Preparation of electrical systems for new buildings to accommodate solar energy harvesting in the future • Inclusion of rainwater harvesting for all new buildings

Source: Green UP Summit, Catalyzing Climate Change Resilience and Environmental Sustainability in 2017 and Beyond 2016.

applications on Integrated Pest Management, and also Integrated Water Management. The project is done in partnership with other state universities and colleges, government agencies, and other institutions.

7.4.2 MODECERA

The Monitoring and Detection of Ecosystems Changes for Enhancing Resiliency and Adaptation in the Philippines (MODECERA) is a DOST-PCAARRD funded project which aims to build an ecological observation network in the country. The ten-year project is geared to provide inputs to promote a science-based approach in the management of the country's natural resources. The project shall also give rise to the Controlled Environment Research Facility (CERF), which will be used to study the response of various natural resources to climate change.

7.4.3 Sustainable Energy Program (SEP)

The Sustainable Energy Program (SEP) is a project of the UP Center for Integrative and Development Studies (UP CIDS) aimed at providing guidance to policy-makers in the creation and reformation of laws that are energy-related. With the Philippines having one of the highest energy prices and fossil fuel consumption per capita worldwide, the project aims to look at this issue. The institute may be divided into four components: Market, Behavior, Technology, and Public Policy. Researches shall be conducted and informations generated are to be disbursed in public lectures, research fora, and the like. One of the Program's component studies is the Pilot Energy Audit for UP Diliman, which determines conservation and efficiency opportunities in the campus through the collection of baseline energy consumption trends. Another is the Life Cycle Benefit-Cost Analysis of Wind Energy in the Philippines, which is aimed at drawing up a quantitative basis for policy recommendations on won energy adoption.

7.4.4 UP DREAM

One of nine (9) component projects of project NOAH, the UP Disaster Risk and Exposure Assessment for Mitigation (DREAM) addresses the prevalence of widespread flooding, deaths, and destruction in the country brought about by typhoons that plague the country each year. The project produces up-to-date high-resolution three-dimensional (3D) maps of the critical basins in the country. The program uses four (4) key components, which are Data Acquisition, Data Validation, Data Processing, and Flood Modeling. Data acquisition is done using light detection and ranging (LiDAR) technology, and the hazard maps produced are then made accessible in the project NOAH website. A more resilient Filipino community is envisioned by the UP DREAM program amidst all environment-related risks.

7.4.5 UPPEG

The UP Program for Environmental Governance (UPPEG) aspires to build the capacity of environmental and natural resources officers of the various local government units (LGUs) in the country. Launched in January 2016, the six-part modular program was crafted with the help of environmental governance experts from UP Los Baños, UP Visayas, and UP Diliman, and is implemented by the UP Center for Integrative and Development Studies (UP CIDS) together with the Department of Environment and Natural Resources (DENR). The program covers a range of topics such as Environmental Governance and the Nature of Environmental Issues in the Philippines, Environmental Leadership and Principles of Conflict Management, and Human, Financial, and Material Resources Management.

7.4.6 NOAH: From Project to Center

Last March 21, 2017, the University of the Philippines, by virtue of the approval of the Board of Regents on February 23 and an Executive Order from UP President Danilo L. Concepcion, established the NOAH Center (Nationwide Operational Assessment of Hazards) – for climate actions and disaster risk reduction management.

The Center takes off from the Department of Science and Technology (DOST)'s research program, Project NOAH, which began in 2012. The past five years saw the project harnessing technologies and management services for disaster risk reduction activities. These were offered by DOST through PAGASA, PHILVOLCS, and the DOST's Advanced Science and Technology Institute (ASTI), in partnership with the UP National Institute of Geological Sciences (NIGS) and the UP College of Engineering. Having delivered its research results and after two extensions, this project under the DOST ended on February 28, 2017.

The University's new NOAH Center will be attached to the Office of the President, with its main office at the UP NIGS in UP Diliman, Quezon City.

The Center aims at "assist[ing] Filipinos in climate change actions and disaster risk reduction by providing timely, reliable and readily accessible data and information, such as hazard risk maps, as a basis for action by warning and response agencies against possible disasters that may occur from floods, typhoons and other natural hazards." As a national scientific research center under UP, the NOAH Center will serve as a reliable and sustainable office tasked with generating "science-based information, models and applications useful for disaster risk reduction and management, climate change adaptation and mitigation, resource management, water conservation and planning, land use and local planning, urban development, and engineering designs, and other similar mainstreaming actions, research, development and extension services."

In addition to its goals of collaboration and partnerships with international and national or local entities or organizations through the UP System, the NOAH Center will also participate in the National Disaster Risk Reduction Management Council's Pre-Disaster Risk Assessment system and provide information, technical assistance and capacity building to the Climate Change Commission.

7.5 Conclusion, Challenges, and the Way Forward

The initiatives of the University show the important role that higher learning institutions play in the transition to a LCE. Clearly, the University continues to take lead role in developing and implementing low carbon initiatives across its functions of instruction, research, extension, and public service. The University has started to reorient its academic operations so as to shape future leaders who will take the competently lead in addressing development concerns such as transition to a LCE. This is evident in the creation of innovative technologies and responsive public service activities.

In recognition of the crucial role of the University's operations in ensuring the transition towards a low carbon campus, the University's strategic plan also emphasizes that operational excellence should go hand in hand with academic excellence. Rudden (2010) states that there is an increasing number of higher education institutions that recognize the necessity the link between financial stewardship and campus sustainability". The University's Green UP initiative is an illustration of what Rudden (2010) describes as the "expanding efforts to identify and optimize the best opportunities for enhancing the sustainability of existing and new campus facilities and infrastructure".

The Green UP initiative, as embodied in the UPMDP document, includes strategic activities such as the implementation of green building design, architectural and engineering standards as well as energy audits. In ensuring operational excellence, the University stays true to its responsibility of leading by example and serving as a living low carbon development model for its partners and the communities and at the same time is able to achieve efficiency in its operations.

Murphy and O'Brien (2014) argue that to "enable colleges and universities to achieve climate neutrality as well as their sustainability goals, institutions need to make a formal commitment, engage all key stakeholders, make strategic multi-faceted decisions tied to specific initiatives and implement institutional change". Although the overarching policy of the University, the UPCMD, was formulated through a highly participatory process, many stakeholders of the University still consider the absence an overall or integrated framework on sustainability in the University. This was raised during the pre-summit consultation for the Green UP Summit held on March 28, 2016. The UPCMD provides for principles and guidelines in the design, building, and operation of the University's physical assets. However, its operationalization is still left with the constituent universities. As such, each unit has its own policies contributing towards the implementation of Green UP, but there are neither integrated guidelines nor policies. Given this, there is also a need to come up with a unified Green UP metrics since just like the policies each unit also uses their own assessment or evaluation tools. Murphy and O'Brien (2014) emphasize the necessity of having "decision-making tools and processes which are practical, affordable and easy to use". One framework that is being proposed by Martin (2011) is the STARs.

Another vital issue that was raised during the consultation workshop was the inadequacy of resources in the form of manpower, budget and even natural resources, i.e., available land spaces, to be able to fully implement the Green UP Program. The University of the Philippines (2016) recognizes the role that higher education institutions play in ensuring inclusive growth and sustainable development thus calls for adequate investments to knowledge capital base.

Despite the challenges faced by the program, the success of Green UP is evident in the numerous activities already being undertaken by the constituent universities. As earlier discussed, each university has implemented programs and projects best suited for them, and some of these best practices may be looked into for possible system-wide replication/adaptation. Some of these projects, in fact, are student-led, such as "Bike Share Program", "Freshmen Native Tree Planting Activity", and the "Exhibit of Green Design Student Project", among others.

The current efforts and knowledge being applied by the constituent universities may even be extended beyond the campus. For instance, a stronger linkage between UP Diliman and its host local government, Quezon City, is highly encouraged, as the exchange knowledge, technology, and best practices may be had. This gives the university an opportunity to take a more pro-active stand and contribute to the community.

References

Carol MacKinnon-Lewis, C., and Frabutt, J. M. (2001). A bridge to healthier families and children: the collaborative process of a university-community partnership. *J. High. Educ. Outreach Engagem.* 6, 65–76.

Dober, R. (1962). *Campus Planning*. Cambridge: Reinhold Publishing Corporation.

Espina, M. A., and Espina, C. S. P. (2013). *Principles of a Sustainable UP Diliman Campus*. Available at: http://www.ovcrd.upd.edu.ph/wp-content/uploads/2013/08/S2_4-ESPINA-NEW-PRINCIPLES-2.pdf

Martin, R. J. (2011). STARS: a campus-wide integrated continuous planning opportunity. *Plan. High. Educ.* 39, 41–47.

Murphy, T., and O'Brien, W. (2014). A strategic decision model for evaluating college and university sustainability investments. *Manage. Res. Rev.* 37, 2–18.

Perry, D. C., and Wiewel, W. (2005). *The University as Urban Developer: Case Studies and Analysis*. Cambridge: Lincoln Institute of Land Policy.

Rudden, M. S. (2010). Five recession-driven strategies for planning and managing campus facilities. *Plan. High. Educ.* 39, 5–17.

Sungu-Eryilmaz, Y. (2009). *Town–Gown Collaboration in Land Use and Development*. Cambridge, MA: Lincoln Institute of Land Policy.

The University of the Philippines (2016b). Knowledge-based development and governance: challenges and recommendations to the duterte administration, 2016-2022. *Int. J. Philipp. Sci. Technol.* 9, 1–9.

Wiewel, W., and Kunst, K. (2007). *University Real Estate Development: Campus Expansion in Urban Settings*. Cambridge, MA: Lincoln Institute of Land Policy.

Documents

Commission on Higher Education (2012). *Roadmap for Public Higher Education Reform*. Available at: http://api.ched.ph/api/v1/download/556

Commission on Higher Education (2017). *2017 Higher Education Facts and Figures*. Available at: http://www.ched.gov.ph/central/page/2017-higher-education-facts-and-figures

University of the Philippines Master Development Plan: Development Principles and Design Guidelines (2014).

The University of the Philippines (2016a). *Green UP Summit: Catalyzing Climate Change Resilience and Environmental Sustainability in 2017 and Beyond*. Quezon City: University of the Philippines.

UP-OVPAA (2014). *Master Development Plan Formulation for Selected State Universities and Colleges (SUCs): A Concept Note*. Quezon City: Office of the Vice President for Academic Affairs.

U.P. Strategic Plan (2011–2017). Available at: https://www.up.edu.ph/wp-content/uploads/2017/05/Stratplan-FINAL-low_res.pdf

8

Determinants of Employees' Perceptions, Commuting Culture, and Environmental Sustainability at Symbiosis International University, India

Vishal Pradhan[1], Saravan Krishnamurthy[1]
and Prakash Rao[2]

[1]Symbiosis Centre for Information Technology, Symbiosis
International University, Pune, Maharashtra, India
[2]Symbiosis Institute of International Business, Symbiosis
International University, Pune, Maharashtra, India

Abstract

This chapter describes a study conducted to develop an understanding of socio-cultural rationalities and behaviour among employee commuters at Symbiosis International University (SIU) in Pune, an Indian metropolis.

Commuting preferences of employees to the university campuses are primarily determined by Cost, Convenience, Comfort and Safety (CCCS) choices. Within particular small groups of employees, these orders of CCCS vary due to an atypical attitude. An 'operational definition of a context' was used to deliberate on possible Willingness to spend, the changes in the behaviour of employees' time, comfort, and safety while commuting to the office. Together with dissatisfactions and importance given to the commuting choices, the Willingness to change commuting patterns has tangible impacts on environmental sustainability. This empirical research assessed the causal structures offered and the interactions due to categories of employees' gender, age, designation, campus location, and vehicle type.

With this backdrop, this social study discusses Pune's metropolis commuting culture of employees. While attaining environmentally sustainable transportation policies for Higher Education Institutes (HEIs) at its heart, this chapter addresses the need for a more integrated understanding of the mitigation challenges and makes rational recommendations for Indian urban HEIs. Employees' willingness and participation in outlined policies would realise the desired behavioural change towards meeting the low-carbon emission goals.

8.1 Introduction

The workplace is a closed social setting. The activities related to once job, even one like commuting to office is influenced by other office mates. At the same time, each office goer is also a part of this collective, and thus influences others. As the commuters are likely to emulate their colleagues, a small change in attitude will reflect in a major change in behaviour (Zaidel, 1992; Redshaw, 2012). The empirical study in this chapter aims to ascertain the groups of office commuters' transportation activities and who would be willing to change their mobility pattern in order to be more environment friendly. The study, assess the causal structure worked on the interaction with employees' age, gender, designation, campus location, and transport mode.

The employee respondents are commuters to Pune (India) campuses of Symbiosis International University (SIU), an urban Higher Education Institution (HEI). A 'premise' is that their commuting preferences are primarily determined by Cost, Convenience, Comfort and Safety (CCCS) motives. For a small number, special groups of employees this order of CCCS and their significances vary due to atypical attitude. For those commuters, the significance of CCCS may have different priorities. The extent of willingness to adopt change by spending, in the commuting pattern has an impact on environmental sustainability (Schade and Schlag, 2003; Devkar et al., 2013). We studied three constructs, namely, Importance, Dissatisfaction, and Willingness. The dissatisfaction and importance weights placed on the CCCS while commuting are accounted as antecedents to Willingness to spend. The possible compromise among CCCS concerns, over low-carbon economy (LCE) status of their HEI, may serve a novel objective in this global warming world, as this study explores.

Towards achieving, the state of LCE via its employees' daily mobility choices to the office, an urban HEI like SIU may explore a 'bottom-up' approach. Before experimenting with (rational) policy implementations, an urban HEI would do a baseline study of transport patterns of its employees', based on a survey design methodology. This would demand intervention for social attitude change (Wells and Beynon, 2011). The evaluation of the conceptual model of daily mobility practices would further give directions for policies operationalisations. Having stake on environmental sustainability, the research aimed to report rational recommendations for the SIU campuses and their employees, eventually for the Urban HEI community in India.

This study has consequently progressed in accord with an exploratory study on carbon footprints of employee commuters of SIU (Rao et al., 2017). Pune urban sprawl daily commuting culture is also a note of this social study. Within the scope, we included observations on how employees perceive and wish to adjust to their commuting to office activities. These were examined from employees' perspective, instead of the perspective of university or society.

8.1.1 India's Responsibility to Environment Sustainability

"One of the fastest emerging economy like India, present interesting dilemmas since rapid mass urbanisation aimed at raising standards of living poses concomitant threats to environmental health ... Increasing evidence of environmental depletion is forcing all concerned to play an active role in improving livability, reducing sustainability burdens and preserving resources for future generations." (Jayanti and Gowda, 2014, pp. 130–142). India is one of the four largest greenhouse gas (GHG) emission countries, referred at the United Nations Framework Convention on Climate Change (UNFCCC 2015) in Paris. In the subsequent agreement, India ratified GHG reduction on global climate goals. While the actual treaty goes into effect from 2020, it is mandatory to start cutting back on carbon emissions. The Ministry of Environment & Forest (MoEF) in partnership with Urban Local Body (ULB), Environment research institutes and HEI, need to be committed to implementing a comprehensive sustainability strategy that protects venerable elements of ecosystem and wellbeing of the citizenry (Revi, 2008; Finlay and Massey, 2012). The time for initiating and managing change has arrived.

8.2 Pune Metropolis Landscape and SIU

Pune was known as the 'bicycle city' of India. With the proliferation of motorised vehicles in the city, 'active mobility' options diminished in proportion (Maunder and Palmner et al., 1997). In an earlier study outlining India's urban transport crisis, Pune's urban trips were compared to other major cities of India with a moderate proportion of commuters on public transport (40% in Pune, 60% in Mumbai, and 80% in Kolkata) (Pucher et al., 2005).

The crisis reported in early 2000 has been growing over the years displaying congestion, pollution and lengthened commute times. Suburban sprawls had begun to exceed limits, given the higher economic growth of the region, aspirations for owning a home, and affordability for vehicles (Basu and Vasudevan, 2013; Paladugula and Rathi, 2013). More than three million vehicles were registered in Pune as of 2016 (RTO, 2016). During the year 2015–2016, a notable statistic, 13,000 two wheelers and 7,000 four wheelers were registered and assumed to be added on the road every year with over 10% CAGR. The Central Institute of Road Transport (CIRT) and Traffic Police have opined that strengthening public transport is the solution. Comparing India and China urban sprawls, the problems arising due to very high consumption of motor vehicles were assessed, and increases in GHG emissions of such urban sprawls were warned, with specifics on swift depletion of non-renewable energy resources (Pucher et al., 2007).

Post-liberalisation of Indian economy in 1992, more than 1,000 industrial units developed in Pune and its suburbia, with a gradually urbanising of nearby villages and expanding resident population densities located at greater distances from Pune's business areas (Dash and Balachandra, 2016). Further growth of businesses and IT revolution made the horizons of the city disappear. Presently, Pune has expanded to be a major tier two city in India but heading towards becoming a megacity in near future. The rapid growth of Pune city has set extra challenge of urban planning for a municipality that did not foresee satellite town development (Dash and Balachandra, 2016; Krishnamurthy et al., 2016). The Pune Municipal Corporation's (PMC's) annual Environmental Status Report includes an analysis of trends or changes in the environment, causes of these changes, assessment, interpretation of implications, and impacts of these trends. The report mentions that the increasing number of vehicles is a major concern for air pollution. It also emphasises on the assessment of actual and potential societal response to environmental problems (PMC, 2016). Among other urban issues, this baseline study pins down on employee commuters' preferences in order to address environmentally sustainable transport.

8.3 Role of HEIs in Environmental Sustainability and SIU

Symbiosis International University campuses in Pune city are part of the Pune city eco-system. HEIs as the centers of knowledge creation need to demonstrate example-setting behaviour in dissemination for sustainable urban transport and relevant GHG reduction (Nuzir and Dewancker, 2014; Schipper et al., 2009). Relevant research, inclusions of environmental sustainability lessons in curricula, and policy development aligned to national targets would impart value based learning (Mulugetta and Urban, 2010). Urban Indian HEIs can achieve citizens' engagement by raising awareness, inducing pro-environmental attitude among its employees and local communities (Aleixo et al., 2017).

Although ULB's role and research institutes expertise are of paramount importance, a proactive step needs to be taken by HEI to help assess environmental depletion, vulnerability, mitigation, and the adaptation strategies (Saveyn et al., 2012; Mulugetta and Urban, 2010). This baseline study would be some take for sustainability scholars that bridge the gap between HEI and the efforts of ULB on best practices in environmental sustainability. This study attempts to predict Willingness to change to more greener commuting practices of SIU employees. In an Indian urban HEI setting, employee commuter's context, if being able to replicate, this study would draw some parallels to global environmental sustainability best practices. Such an effort would require a long-term focus on institutionalising sustainability within the HEI. In India, although, providing critical insights into the success of sustainability strategies is still an ongoing work.

8.4 Urban HEI Commuting Culture, Attitude, and Changes

Pune ULB jurisdiction share boundary with one of the largest populated manufacturing firms' hubs, Pimpri-Chinchwad ULB. Pune, traditionally known as the 'Oxford of the East' also expanded in establishing new private HEIs, attracting about 20% of all international students bounded for India. As a growing educational hub, the responsibility of LCE encompasses several organisations across sectors, ranging from manufacturing to services and HEIs. To decelerate GHG emissions, changes in fast-urbanising world populations are required, especially in million plus cities (Fatima and Kumar, 2014). Indian urban HEIs need to gather valuable baseline data and attempt to develop solutions active LCE-based thought to develop local sustainability projects with learned collaborations. Some infrastructural additions such as City Bus Rapid Transit assisted in regulating traffic congestions (Kathuria et al., 2016). It is currently impracticable to demand more from the civil authorities struggling with project overruns (Vaidyanathan et al., 2013). Opposing, shortcomings were noted in easing congestions or GHG control. For instance, less than 3% of Pune city roads have bicycle tracks (PMC, 2012). The willingness of citizens' participation in upgrading their walk and bicycle tracks is illustrated in the section below.

8.4.1 Willingness to Spend Aspect

A World Bank study in 2008 aimed at urban India, proposed a project to improve non-motorised transport for Pune people (Wang et al., 2011). This Sustainable Urban Transport Program (SUTP) planned to improve active transportation. The contingent valuation method implemented assessed that citizen's Willingness to spend was lower than required project cost. In contrast, charismatic individual championing could produce proactive participation of citizens and Willingness to spend, if the cause was as undeniable as municipality waste management (Devkar et al., 2013). These examples attempt to build a referendum highlight the need to study Willingness to spend. Another empirical study on bus transit system in the Indian city of Bardoli, Gujarat found encouraging results on cost savings, commuting time-saving, efficiency, and willingness for shifting to bus transit (Fatima and Kumar, 2014). Although sustainability issues were well addressed, the attitudinal aspect of commuters' willingness requires an in-depth study, with emphasis on the culture and ethics of the organisation (Jayanti and Gowda, 2014). Research on transportation psychology elucidates that acceptability of public transport pricing strategies is expectedly low (Schade and Schlag, 2003). Acceptability was positively related to the social norm, personal outcome expectations, and perceived effectiveness. The high awareness of the problems will lead to increased willingness for accepting solutions. Further, the need for empirical studies on the influence of problem perceptions is highlighted, since they were generally inconsistent. We argue that the commute related problems such as costs and environmental problems are essential in the current study. The current study limits itself from multi-level perspectives where ecologically benign socio-technical practices and uninformed stakeholders may exist (Smith et al., 2010). However, gathering significant insights for sustainability transitions by focusing on the situated potentials of multi-level processes within an HEI commuter's world is noted as input for current research (Smith et al., 2010; Vaidyanathan et al., 2013).

8.4.2 Dissatisfaction Aspect

Demand Responsive Transport (DRT) emerged as the middle way of transport instead of the cumbersome, nonflexible public transport and expensive, self-driven personal vehicle. (Davison, et al., 2012). In the UK, high customer satisfaction levels were observed but a lack of frequency was noted. In Pune, the DRT provided by SIU in common bus routes of employees at a subsidised cost. This mode does not have a novelty barrier (Davison et al., 2012).

At the operational level, only employees of SIU were eligible for transport, with adequate responsiveness to social and economic concerns, as gathered by survey responses. Geographical coverage was found to be adequately satisfactory for most commuters. However, the appreciation for continued services and the opportunity to unwind in the company of co-workers during the return to home commute was expressed. While socio-cultural bonds may form at the micro-level of the individual employee, a socio-cultural shift is required at the consumer's group level. Further open discussions and stakeholder engagement in decision making is essential for improvement of commuting services aimed towards LCE.

8.5 Methods and Validations

8.5.1 The Study Area of This Empirical Research

SIU is a progressive HEI with seven campus locations in Pune (Figure 8.1). *S.B Road* is its first campus within urban city limits but was distant from the inner city. With the demand for professional programs and quality education other SIU campuses started. *Atur centre*, *Model colony*, and *Khadki* campuses are in close vicinity of *S.B. Road* campus. In this study, we are considering them together as a Main city campus (Ref. Green pointers). The *Viman-Nagar* and *Hinjawadi* campuses were established beyond urban limits. Today these two campuses are part of the much developed urban suburbs. In this study,

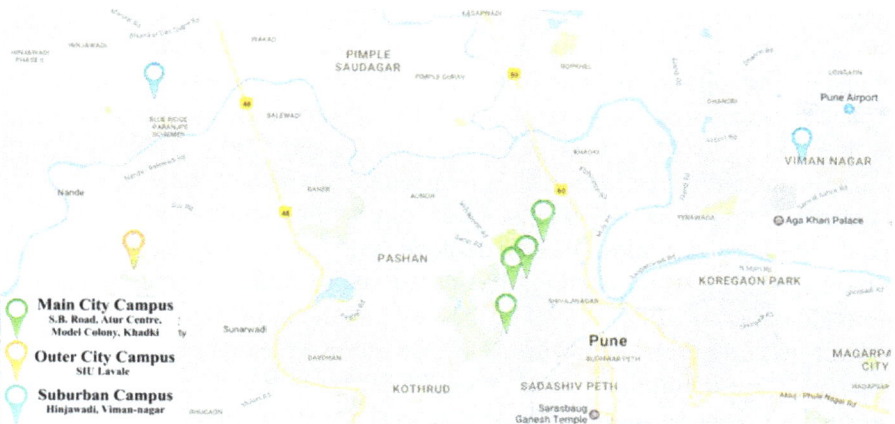

Figure 8.1   Symbiosis International University campuses in Pune metropolis.

Source: Google Maps © 2017.

we are considering them as the Suburban Campus (Ref. Cyan pointers). The SIU, *Lavale* campus is located way outside Pune urban limits (Ref. Orange pointer).

8.5.2 The Questionnaire

A structured online and/or guided survey was administered to collect information on SIU employees' perceptions on commuting to their respective campuses. The employees who usually commute during peak traffic hours (before 9 am and after 5 pm), belonged to various typologies, (comprised of Teaching Staff, Management employees of junior, middle and senior cadres of various Age, Socio-Cultural-Economic background, and Gender). In the survey questionnaire, Section-A included the demographic questions and typology information. Section-B recorded the perception of the employees about their commuting activities. Using the standard five-point Likert scale, the questions were aimed at capturing employees' opinions and decisions about their commuting practices. Section-C aimed at determining the extent of the commuting choices' influence on employees' perception regarding CCCS, using a five-point Likert scale. These were the indicators of Importance, Dissatisfaction, and the behavioural change implications, namely, Willingness to spend. The data transcribed was further screened for data fitness prior to statistical tests. The descriptive analysis also indicated to do detailed data analysis. 360 responses were received after the survey, out of which 355 were retained after the logical inspection and data cleaning.

The scales for Importance and Dissatisfaction factors were constructed by understanding 'operational definition of the context' (Zimmermann et al., 2007). These were operationally defined as a linear sum of the measurements on the descriptions of Cost, Convenience, Comfort, and Safety (Table 8.1). Accordingly, the employee respondents mapped the domain of these factors. The contexts of Importance, Dissatisfaction and Willingness to spend were pre-fixed sections of the questionnaire deployed and based on the qualitative panel inputs. As they were based on the operational definition characterising functions (Zimmermann et al., 2007), the issue of the construct validity of the contexts was naturally satisfied. Also this survey research comment on the gravity of the common method bias (CMB) by noting that the information on CCCS variables was concerned to the respondents themselves. In addition, the proactive design step to mitigate threats of method effect was taken (Conway and Lance, 2010), by means of operational definition

Table 8.1 Describing the factors, test for loading >0.4 and quality criterions

Factor Label	Variables' Label	Factor Loadings	Quality Criterions
Dissatisfaction with commuting issues	Dissatisfaction with commuting convenience	0.806	Extraction SS loadings 30.196%,
	Dissatisfaction with mode of transport	0.777	
	Dissatisfaction with high costs of commuting	0.760	
	Dissatisfaction with physical safety	0.718	
	The unhappiness that demand self to must make changes in commuting routines.	0.583	Cronbach α is .862
Willingness to spend, the changes in behaviour	Will spend more for safety	0.923	Extraction SS loadings 13.828%,
	Will spend more to save time	0.826	
	Will spend more for Comfort	0.771	Cronbach α is .884
Importance of commuting choices	Importance of convenience	0.841	Extraction SS loadings 10.895%,
	Deciding on type, Importance of safety	0.768	
	Importance of access to mode	0.750	Cronbach α is .768
	Importance of cost of commute	0.400	
			TVE = 54.92 %

Source: Authors.

of contexts. The above stated 'premise' that commuting preferences are dependent on CCCS concerns and qualitative inputs from the respondents annulled the effect of overlap in items for Importance, Dissatisfaction, and Willingness constructs. The expected changes in the behaviour were rectified in terms of Safety, Time, Comfort of the Willingness to spend factor.

8.5.3 Validation of the Contexts

Exploratory factor analysis (EFA) was a recommended technique when dealing with a group of inter-correlated variables, which represent some common aspects of the underlying topic of discussion. The goal of factor analysis was to minimise the number of dimensions. EFA was conducted on 355 responses with 13 observed variables. Three factors were drawn using maximum likelihood factor analysis (MLFA) method (Harman, 1976; Morrison, 1998). The need for doing factor analysis was additionally to validate whether the extracted factors were in congruence with the sections of the questionnaire instrument. It was noteworthy that the questionnaire sections and factors evolved due to MLFA were equivalent. The fitness of the analysis strictly adhered. The Promax rotation method with Kaiser normalisation was used as the factors were assumed correlated. The factors were interpreted as scales of Importance to CCCS, Dissatisfaction to CCCS and Willingness to spend from the factor loading matrix, after suppressing values less than .4. The factor descriptive labels were ascertained after the interpretation of the variables loaded on in each of the three factors. Dissatisfied with convenience, commute mode, comfort, costs, safety, were indicators of Dissatisfaction. The importance given to convenience, Importance of mode of commuting, lower costs, safety, can have a significant impact on the variables that define the Importance factor. Also, Willingness to spend is a specific change in behaviour factor.

8.5.3.1 Reliability analysis for internal structure of tests

The overall Cronbach (1951) α is .862 (5 items), .884 (3 items), .768 (4 items) for each of the sub-scale of Dissatisfaction, Willingness to spend and Importance factors respectively. In addition, none of the items on a sub-scale would increase the reliability if they were deleted. This indicates that all items are positively contributing to the overall reliability. The overall α is excellent because it is above 0.7, and indicates good reliability (Tabachnick and Fidell, 2007).

8.5.4 Posited Structural Model and Hypotheses

We proposed a factor structure based on the presumption that Importance may also have an indirect impact on Willingness to spend—for more safety, for saving more time, for more comfort, via the intervening Dissatisfaction factor (Figure 8.2). We also took clues from the factor scores correlations to roughly ascertain the model. The suggested simple mediation model was tested using sequential regression modeling, PROCESS tool (Hayes, 2013) enabled in IBM-SPSS (Table 8.1).

We tested the significance of indirect effect hypothesis:

H1: The perceived Importance given to CCCS is a significant predictor of Willingness to spend, the changes in behaviour, due to the mediation effect of the Dissatisfaction due to CCCS reasons.

The subsequent hypotheses were tested using causal step approach (Baron and Kenny, 1986) (Table 8.2).

H2: The Importance has a significant total impact on the Willingness to spend (Figure 8.3).

H3: The Importance has significant direct impact on the Dissatisfaction Factor.

H4: The overall effect of the Importance and the Dissatisfaction on the Willingness to spend is significant.

H4a: The effect of the Dissatisfaction on the Willingness to spend is significant, controlling for the effect of the Importance.

H4b: The effect of the Importance on the Willingness to spend is significant, controlling for the effect of the Dissatisfaction.

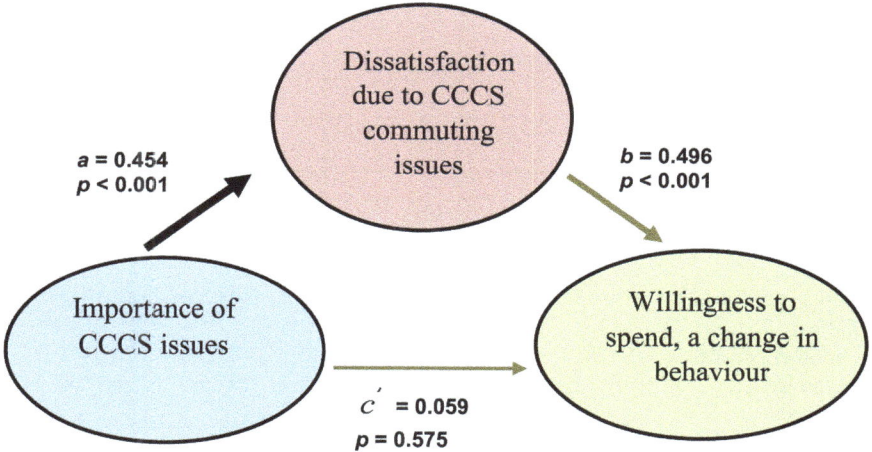

Figure 8.2 Proposed Model based on Factor Analysis. Mediator = Dissatisfaction; Focal Predictor = Importance; Response and variable = Willingness to spend.

Source: Authors.

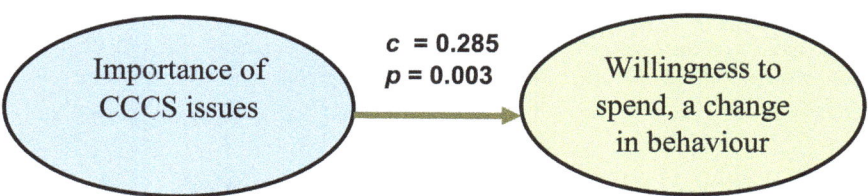

Figure 8.3 Total Effect Model. Focal Predictor = Importance; Response variable = Willingness to spend.

Source: Authors.

Table 8.2   Regression results for the mediation of the effect of importance on willingness to spend by dissatisfaction (Table format: Preacher and Kelly, 2011)

Model	Estimate of Effect Size	SE/ (Boot-Strap SE)	p-Value	LLCI/ (Boot-Strap LLCI)	ULCI/ (Boot-Strap ULCI)
Total Effect Model (without mediator) (Figure 8.3)					
Intercept	1.716	–	–	–	–
Importance → Willingness to spend	$c = 0.285$	0.096	0.003	0.095	0.474
$R^2{}_{Y,X}$	0.020	–	0.003	–	–
Model with Mediator (Figure 8.2)					
Intercept	1.251	–	–	–	
H3: Importance→ Dissatisfaction	$a = 0.454$	0.126	<0.001	0.207	0.701
H4a: Dissatisfaction → Willingness to spend	$b = 0.496$	0.060	<0.001	0.379	0.614
H4b: Importance → Willingness to spend	$c' = 0.059$	0.105	0.575	–0.148	0.266
Indirect Effect ($a \times b$)					
Unstandardised	0.226	(0.070)	<0.01	(0.103)	(0.377)
Partially standardized indirect effect	0.192	(0.058)	<0.01	(0.088)	(0.314)
Completely standardised indirect effect	0.112	(0.031)	<0.01	(0.055)	(0.174)
R^2 mediation effect size	0.019	(0.012)	<0.01	(0.000)	(0.046)
Preacher and Kelley Kappa-squared	0.117	(0.031)	<0.01	(0.058)	(0.180)
Normal theory test	0.226	0.068	0.001	–	–
$R^2{}_{M,X}$	0.059	–	<0.001	–	–
$R^2{}_{Y,MX}$	0.220	–	<0.001	–	–

Regression weights a, b, c, and c' are illustrated above. $R^2{}_{Y,X}$ is the proportion of variance in Y explained by X, $R^2{}_{M,X}$ variance in M explained by X, and $R^2{}_{Y,MX}$ is the proportion of variance in Y explained by X and M. The 95% CI for $a \times b$ is obtained by the bias-corrected bootstrap with 1,000 resamples. Here, Importance is the independent variable (X), Dissatisfaction is the mediator (M), and Willingness to spend is the outcome (Y). CI (lower) = lower bound of a 95% confidence interval; CI (upper) = upper bound; = direction of effect. LLCI/ULCI is Lower/Upper limit of confidence interval. *Source:* Authors.

8.5.4.1 Assessment of the proposed mediation model

The MLFA incorporated the factor structure that followed a series of mediation analyses (Tables 8.2 and 8.3). We tested the significance of a series of relations (**H2, H3, H4, H4a,** and **H4b**) between the factors shown within the ellipses before specifically testing **H1** using causal step approach

Table 8.3 Mediation model effect sizes for the typologies of commuters

Typologies, (N of Commuters)	Direct Effect Size, p-Value	κ^2-Indirect Effect (Size[a]), 95% Bootstrap CI	Total Effect Size, p-Value
Gender			
Female, (N = 146)	$c' = -0.199$, $p = 0.185$	$\kappa^2 = 0.191$ (Large), [.107, 0.286]	$c = 0.159$, $p = 0.357$
Male, (N = 209)	$c' = 0.213$, $p = 0.092$	$\kappa^2 = 0.086$ (Medium), [0.017, 0.166]	$c = 0.378$, $p = 0.005$
Age			
<30 years, (N = 94)	$c' = 0.278$, $p = 0.302$	$\kappa^2 = 0.178$ (Medium), [0.053, 0.349]	$c = 0.787$, $p = 0.01$
30–40 years, (N = 175)	$c' = -0.028$, $p = 0.83$	$\kappa^2 = 0.117$ (Medium), [0.056, 0.197]	$c = 0.181$, $p = 0.193$
> 40 years, (N = 86)	$c' = 0.113$, $p = 0.515$	$\kappa^2 = 0.069$ (Small), [0.004, 0.221]	$c = 0.227$, $p = 0.226$
Designation			
Teaching, (N = 127)	$c' = 0.071$, $p = 0.641$	$\kappa^2 = 0.015$ (Small), [0.000, 0.064]	$c = 0.096$, $p = 0.549$
Administration (N = 46)	$c' = 0.131$, $p = 0.657$	$\kappa^2 = 0.182$ (Medium), [0.018, 0.37]	$c = 0.53$, $p = 0.163$
Non-teaching (N = 182)	$c' = -0.018$, $p = 0.904$	$\kappa^2 = 0.163$ (Medium), [0.095, 0.239]	$c = 0.324$, $p = 0.038$
Campus Location			
Main City, (N = 127)	$c' = -0.066$, $p = 0.647$	$\kappa^2 = 0.124$ (Medium), [0.046, 0.231]	$c = 0.142$, $p = 0.386$
Suburban, (N = 96)	$c' = 0.196$, $p = 0.341$	$\kappa^2 = 0.089$ (Medium), [0.029, 0.186]	$c = 0.378$, $p = 0.062$
Outer City, (N = 132)	$c' = 0.178$, $p = 0.302$	$\kappa^2 = 0.019$ (Small), [0.001, 0.098]	$c = 0.215$, $p = 0.202$
Mode of Commute			
Campus Bus, (N = 150)	$c' = 0.051$, $p = 0.774$	$\kappa^2 = 0.169$ (Medium), [0.082, 0.262]	$c = 0.436$, $p = 0.011$
Car, (N = 54)	$c' = -0.249$, $p = 0.388$	$\kappa^2 = 0.137$ (Medium), [0.013, 0.355]	$c = 0.036$, $p = 0.914$
Bike, (N = 136)	$c' = 0.121$, $p = 0.409$	$\kappa^2 = 0.042$ (Small), [0.001, 0.136]	$c = 0.196$, $p = 0.247$

Number of bootstrap samples for bias-corrected bootstrap CI is 5000. The level of confidence for all CI in output is 95%.

[a]The heuristic limits on the κ^2 mediation effect sizes values, as small (0.01), medium (0.09) and large (0.25) (Cohen, 1988).

Source: Authors.

(Baron and Kenny, 1986) (Figures 8.2 and 8.3). The various relation strengths (effect sizes and their significances) were shown on the path diagrams with appropriately thickened arrows. Analytically calculated confidence interval (CI) estimates of the estimated effect sizes were reported. Regression results for the effect of Importance on CCCS on Willingness to spend on CCCS mediated by Dissatisfaction were tabulated for brevity (Table 8.2). In general, a hypothesis test for the effect size is significant if the calculated 95% CI does not include hypothesised zero effect size (and $p < 0.05$).

The posited total effect model (Figure 8.3) explains the key relationships of the Importance placed on CCCS as perceived by the employees on the Willingness to spend, the changes in behaviour regarding CCCS was significant, $c = 0.285$ ($p = 0.003$, 95% CI [0.095, 0.474]). Thus, the test for **H2** holds sufficiency condition to test model in Figure 8.2. This significance emphasises the relation between the Importance of CCCS on the Willingness to spend without Dissatisfaction factor. The relation of Importance of CCCS on the Dissatisfaction due to CCCS has the significant positive effect, $a = 0.454$ ($p < 0.001$, 95% CI [0.207, 0.701]) (Figure 8.2). Thus, the test for **H3** was favourable. Employees' feel that CCCS considerations are important, in turn, makes them feel about the Dissatisfaction due to CCCS. The (overall) effect of Importance of CCCS and Dissatisfaction due to CCCS on the predictive ability of Willingness to spend was significant ($F = 39.993$, $p < 0.001$). Thus, the test for **H4** was found favourable. The Dissatisfaction due to CCCS was found partially significant predictor of Willingness to spend, with the effect size $b = 0.496$ ($t = 8.323$, $p < 0.001$, 95% CI [0.379, 0.614]) (Figure 8.2), controlling for the effect of Importance of CCCS. Thus, the test for **H4a** was found favourable. The effect of Importance of CCCS was found partially non-significant predictor of Willingness to spend, with the effect size $c' = 0.059$ ($t = 0.561$, $p = 0.575$, 95% CI [–0.148, 0.266]) (Figure 8.2), controlling for the effect of Dissatisfaction due to CCCS. Thus, the test for **H4b** was not rejected. Thus, the total effect model (Figure 8.3) after the inclusion of mediator become evidence of a nonsignificant direct relation between the Importance on the Willingness to spend factor. Thus, Figure 8.2 depicted a 'complete' mediation model.

The overall predictability of the employees' Willingness to spend, to improve their concerns due to CCCS issues is mainly affected by their Dissatisfaction due to CCCS reasons than the weight they place on the Importance of CCCS issues. The SIU's overall DRT service can attempt to minimise the Dissatisfaction with commuting experiences due to CCCS issues. The Importance of CCCS is an antecedent to the Dissatisfaction due to CCCS. If SIU employees give Importance to CCCS along with Dissatisfaction due to CCCS, then they can positively induce initiations for Willingness to spend, the changes for pro-environmental behaviour.

Notably, the significance of effect size $b = 0.496$ is nearly same of the effect size $a = 0.454$. This means that the effect of Dissatisfaction due to CCCS on Willingness to spend likely may be strong/weak if perceived effect of Importance of CCCS on Dissatisfaction increase/decrease. Thus, rational efforts to nullify commuting Dissatisfaction due to CCCS concerns among employees' and strategies are antecedent to the extent of Importance placed on CCCS issues while commuting. The Indian urban HEI who provide DRT to their employees may take this point.

8.5.4.2 Assessment of indirect effect size measures of the commuting dissatisfaction

Analysis of indirect effect is most important in mediation analysis. Preacher and Kelley (2011) have suggested using more than one effect size measures based on their meaningfulness of metric, construction of confidence interval (CI), and independence of sample size issues. This approach would overcome fallacies of causal step methodology by Baron and Kenny (1986) (Hayes, 2009, 2013). Specifically, we tested H1. In general, a hypothesis test for the indirect effect size is significant if the calculated 95% Bootstrap CI does not include hypothesised zero effect size. The assessment of indirect effect sizes using various measures, such as; Unstandardised, Partially standardised, Completely standardised and R^2 mediation indirect effect sizes, confirmed the significance of mediation effect (Table 8.2).

The most recommended (confirmatory) indirect effect size measure is κ^2 (kappa-squared), which is free from the original scale of the variables. κ^2 is interpreted as the proportion of the maximum possible indirect effect that could have occurred, provided the constituent effects are as large as design and data permit (Preacher and Kelly, 2011). Cohen, 1988 proposed heuristic limits on the κ^2 mediation effect sizes values, as small (0.01), medium (0.09) and large (0.25). Accordingly, $\kappa^2 = 0.117$ (95% Bootstrap CI [0.058, 0.180]), the mediation effect is larger than small (because lower limit of Bootstrap CI excludes 0.01), but smaller than large (because the upper limit of Bootstrap CI excludes 0.25). Thus, it is safe to report that mediation effect size is 'Medium', apart from being 'Complete' mediation model in Figure 8.2. The prediction of the Willingness to spend was found significant in all the indirect effect size tests thus reported.

8.5.5 Mediation Assessments for Commuters' Typologies

We did mediation tests on the above set of hypotheses that study Willingness to spend behaviour by incorporating categories of commuters' typologies

(Table 8.3). The tests for direct, indirect and total effects derive interesting empirical interpretations as discussed further.

8.5.6 Results and Interpretations of the Survey

Among all the indicator variables (Table 8.1), the one which are greater than 0.7 loading onto the factors were considered of having greater indicator reliability (Hair et al., 2014). First, the Importance about Safety during the commute (0.768) had strongest reliability on the importance factor. The second among remaining variables, the employees value 'convenience of commuting'. Dissatisfaction with convenience (0.806), exhibited strongest indicator reliability on Dissatisfaction factor. Lastly, employees' value Saving time and Comfort, as personal Safety (0.923), Saving time (0.826) and Comfort (0.771) had the strongest indicator reliabilities on Willingness to spend factor. Dissatisfaction indicator of self-compulsion to make changes in commute pattern (0.583) and Importance indicator of cost (0.400) had the weak factor loadings. The voluntary compulsion on himself/herself for making changes to fix Dissatisfaction experience of commute and weak motive of Importance of spending money can put doubt on overall employees' Willingness to spend for improving their CCCS concerns. This order determines the determinants of environmental sustainability study of an Indian urban HEI.

The interpretations assume ceteris paribus conditions and are based on the typology of employee commuters. By typology, the employees place Dissatisfaction on CCCS issues. The medium or large significance of the indirect effect of the Importance on Willingness to spend by Dissatisfaction was observed in many cases of the commuter type. Further, the feeling of Dissatisfaction with present commuting arrangement would notes the change in behaviour by making them likely to spend by will, which minimise the concern of CCCS issues. The Importance given to CCCS issues by the employee commuters would almost always not bring in the behavioural change of willfully spending to improve the sustainable commuting experience.

Regarding Male commuters ($N = 209$), the total effect (the sum of direct and indirect effects via two causal paths from Importance to Willingness) is significant hinting towards possibility in Male commuters for accepting Willingness to change by spending ($c = 0.378$, $p = 0.005$). The Male employees may be counseled, for Willingness to spend as they also perceive near significant Importance to improve CCCS led commuting experience

$(c' = 0.213, p = 0.092)$. For the commuters younger than 30 years ($N = 94$), the total effect is significant cueing towards likelihood in the commuters younger than 30 years for accepting Willingness to change by spending ($c = 0.787, p = 0.01$). Non-teaching service staff ($N = 180$), the total effect is significant hinting towards prospect in non-teaching service staffers for accepting Willingness to change by spending ($c = 0.324, p = 0.038$). Suburban campus employee ($N = 96$) may be counseled to improve their agreement on Willingness to spend to improve CCCS led commuting experience as this test is marginally non-significant ($c = 0.378, p = 0.062$). Campus bus commuters ($N = 150$), the total effect is substantial ($c = 0.436, p = 0.011$) inferring the chance in campus bus commuters for accepting Willingness to change by spending to better their CCCS issues. Thus, Males, Younger commuters less than 30 years, Non-teaching service staff, Suburban campus employee and Bus commuters may be sensitised to realise pro-environmental cultural change in behaviour before exercising policies (Wells and Beynon, 2011).

Conversely, for employees greater than forty years age ($N = 86$), Female employees ($N = 146$), Teaching staff ($N = 127$), Main city campus commuters ($N = 127$), Outer city campus ($N = 132$) and solo Car commuters ($N = 54$) are more 'rigid' towards their office commuting preferences and practices. Referring to Indian social set-up female employees (Levy, 2013), though they are somewhat Dissatisfied with their commute are likely to have rigid commuting practices. Main city campus commuting practices of the employees need not be 'over fitted', as their commute is already much environmentally sustainable (Rao et al., 2017). Also, Outer city campus commuting practices of employees are 'tricky' to fit into sustainable commuting norms, may be due to the distance of commute. For various combinations of remaining employees (older than forty years age, Teaching staff, and solo Car commuters) we recommend extensive vehicle sharing, enabled by ICT and adopting private mass rapid transport (Rao et al., 2017). In addition, surveyed qualitative inputs encourages discussions to explore carbon economical options of telecommuting, staggering office hours to avoid traffic congestion, and adopt active mobility modes.

8.5.7 Study Outcomes and Discussion

We applied qualitative interpretations along with mediation analyses. Dissatisfaction reasons act as a mediator between CCCS facets of Importance and its impact on Willingness to spend for desirable change measures to address environmental sustainability subsequently. The effect of Dissatisfaction on Willingness to change is likely to become redundant if perceived effect of Importance on Dissatisfaction diminishes. We aspired to delve into 'how' commuting to a university job could positively influence employee well-being and enhance relevant environment friendly sustainability practices. Such analyses provided a scientific approach to decision making for urban Indian HEI, government bodies and policy makers (Finlay and Massey, 2012). This study may be a useful empirical role model for other urban HEI in India. HEIs need to develop a partnering frame of mind with their employees for administrating effective commuting policy (Wells and Beynon, 2011).

An urban HEI practicing sustainable commuting would be perceived with high rewards with least GHG emission. The sense of employee belongingness with HEI would enhance if they are well informed about envisioned sustainability goals and activities. Employees are likely to identify, engage better and demonstrate Willingness to spend behaviour. Additionally, employees may influence their HEI to initiate robust and sustainable commuting practices. Employees may express wishes for the HEI to make their office commuting policy mutually useful (Wells and Beynon, 2011). Successful implementation of policies are likely to happen when HEIs follow sustainable and mutually useful commuting practices. Employees and HEIs need to work together to gain clarity in achieving sustainability goals holistically.

8.5.7.1 Scope for future research

The individualistic-collectivistic nature of commuting culture of office goers (Ziadel, 1990) can achieve environment-friendly commuting practices. A unique early persuasion strategy may be experimented with its success to be realised in the future. The change may be escalated to the larger group of commuters, making it an interesting proposition for the future research (Redshaw, 2012). The improved culture of commuting to address the social issue of environmental sustainability may be assessed using an empirical model. The employees' willingness of participation in outlined policies would realise the success of desired behavioural change.

We suggest to include appropriate more mediators to obtain larger mediation effects and test a more comprehensive model. A series of studies may be undertaken to study various campus sustainability issues, regarding water, electricity, Sewage treatment, bio-degradable waste management (food waste generated in students' bulk kitchen, and garden maintenance), recycling resources and use of clean options, etc.

The findings of this empirical study may be generalised to the Indian Urban HEI's sustainable commuting policies that have larger implication on the environment. The researchers may empirically recommend, based on the employees' observed reparation on the outcome of Willingness to spend, the HEI's efforts to implement baseline policies for environmentally sustainable campus commuting. Environment sustainability may be perceived more as interference instead of a wholesome engagement with society by an urban HEI. Simultaneously, we qualitatively noted that employees happy with their commuting found fewer stresses and better identify with their HEI; but within the scope of this study did not infer in clear quantitative terms 'how they achieved this balance'. These broad research queries may lead further to derive a clear scope on the validation of the longitudinal study determinants of HEI's sustainability commuting practices and its influence on employee perception'.

8.6 Conclusion

This chapter attempts to develop a better understanding regarding employees commuting preferences for an Urban HEI in India. Empirical evidence points out the Willingness to spend and possibilities of change among commuters, based on campus location, gender, designation and age of the employee. This study attempts to set a scientific basis for many HEIs across India in understanding their employees' commute preferences. It is clear that HEIs can contribute towards a much healthier environment by adopting proper sustainability practices. Including commuting patterns within sustainability practices is a useful example that directly instills sustainability inspirations among HEI administrations and educates HEI employees. If adopted in a widespread manner, such practices of surveying employees and encouraging commute changes are likely to support further discussions and similar examples across all departments to perform with sustainability thinking.

Acknowledgement

This research project was funded by a SIU minor research grant. The authors gratefully acknowledge SIU, the support extended by several departments during data collection, and the various Symbiosis employees for their multiple inputs in developing this study.

References

Aleixo, A. M., Azeiteiro, U. M., and Leal, S. (2017). "UN decade of education for sustainable development: perceptions of higher education institution's stakeholders," in *Handbook of Theory and Practice of Sustainable Development in Higher Education*, eds L. Filho, W. Azeiteiro, U. M. Alves, and Fatima (Cham: Springer International Publishing), 417–428.

Baron, R. M., and Kenny, D. A. (1986). The moderator–mediator variable distinction in social psychological research: Conceptual, strategic, and statistical considerations. *J. Pers. Soc. Psychol.* 51:1173.

Basu, S., and Vasudevan, V. (2013). Effect of bicycle friendly roadway infrastructure on bicycling activities in urban India. *Procedia* 104, 1139–1148. doi: 10.1016/j.sbspro.2013.11.210

Cohen, J. (1988). *Statistical Power Analysis for the Behavioral Science*. New York, NY: Academic Press.

Conway, J. M., and Lance, C. E. (2010). What reviewers should expect from authors regarding common method bias in organizational research. *J. Bus. Psychol.* 25, 325–334.

Cronbach, L. J. (1951). Coefficient alpha and the internal structure of tests. *Psychometrika* 16, 297–334.

Dash, N., and Balachandra, P. (2016). Benchmarking urban sustainable efficiency: a case of Indian cities. *Transp. Res. Proced.* 14, 1809–1818. doi: 10.1016/j.trpro.2016.05.147

Davison, L., Enoch, M., Ryley, T., Quddus, M., and Wang, C. (2012). Identifying potential market niches for Demand Responsive Transport. *Res. Transp. Bus. Manage.* 3, 50–61. doi: 10.1016/j.rtbm.2012.04.007

Devkar, G. A., Mahalingam, A., and Kalidindi, S. N. (2013). Competencies and Urban Public Private Partnership projects in India: a case study analysis. *Policy Soc.* 32, 125–142. doi: 10.1016/j.polsoc.2013.05.001

Fatima, E., and Kumar, R. (2014). The introduction of public bus transit in Indian cities. *Int. J. Sustain. Built Environ.* 3, 27–34. doi: 10.1016/j.ijsbe.2014.06.001

Finlay, J., and Massey, J. (2012). Eco-campus: applying the ecocity model to develop green university and college campuses. *Int. J. Sustain. High. Educ.* 13, 150–165. doi: 10.1108/14676371211211836

Hair, J. F., Black, W. C., Babin, B. J., Anderson, R. E., and Tatham, R. L. (2014). *Multivariate Data Analysis*, 7th Edn. New Jersey: Prentice Hall.

Harman, H. H. (1976). *Modern Factor Analysis.* Chicago, IL: University of Chicago Press.

Hayes A. F. (2013). *Introduction to Mediation, Moderation, and Conditional Process Analysis: A Regression-Based Approach.* New York, NY: Guilford Press.

Hayes A. F. (2009) Beyond Baron and Kenny: Statistical mediation analysis in the new millennium. *Commun. Monogr.* 76, 408–420.

Jayanti, R. K., and Rajeev Gowda, M. V. (2014). Sustainability dilemmas in emerging economies. *IIMB Manage. Rev.* 26, 130–142. doi: 10.1016/j.iimb.2014.03.004

Kathuria, A., Parida, M., Ravi Sekhar, C., and Sharma, A. (2016). A review of bus rapid transit implementation in India. *Cogent Eng.* 3:1241168. doi: 10.1080/23311916.2016.1241168

Krishnamurthy, R., Mishra, R., and Desouza, K. C. (2016). City Profile: Pune, India. *Cities* 53, 98109. doi: 10.1016/j.cities.2016.01.011

Levy, C. (2013). Travel choice reframed: "deep distribution" and gender in urban transport. *Environ. Urban.* 25, 47–63.

Maunder, D., Palmner, C., Astrop, A., and Babu, M., (1997). "Attitudes and travel behaviour of residents in Pune, India attitudes and travel behaviour of residents in Pune, India," *Proceedings of the 76th Annual Meeting, Transportation Research Board,* January 12–17, Washington, DC.

Morrison, D. F. (1998). *Multivariate Analysis, Overview.* New York, NY: John Wiley & Sons, Ltd.

Mulugetta, Y., and Urban, F. (2010). Deliberating on low carbon development. *Energy Policy* 38, 7546–7549.

Nuzir, F. A., and Dewancker, B. J. (2014). Understanding the role of education facilities in sustainable urban development: a case study of KSRP, Kitakyushu, Japan. *Proced. Environ. Sci.* 20, 632–641. doi: 10.1016/j.proenv.2014.03.076

Paladugula, A. L., and Rathi, S. (2013). Strategies to reduce energy use for commuting by employees. *Procedia* 104, 952–961. doi: 10.1016/j.sbspro.2013.11.190

PMC (2012). *Revised City Development Plan for Pune.* Available at: http://punecorporation.org/informpdf/CDP/Executive_Summary-Revised%20CDP.pdf

PMC (2016). Available at: https://pmc.gov.in/en/esr-report-2015-2016.

Preacher K. J., and Kelley K. (2011). Effect size measures for mediation models: Quantitative strategies for communicating indirect effects. *Psychol. Methods* 16:93.

Pucher, J., Korattyswaropam, N., Mittal, N., and Ittyerah, N. (2005). Urban transport crisis in India. *Transp. Policy* 12, 185–198.

Pucher, J., Peng, Z. R., Mittal, N., Zhu, Y., and Korattyswaroopam, N. (2007). Urban transport trends and policies in China and India: impacts of rapid economic growth. *Transp. Rev.* 27, 379–410.

Rao, P., Krishnamurthy S., and Pradhan, V. S. (2017). Commuters' carbon footprints – a sustainability case study from symbiosis international university, India. Higher Education Institutions in a Global Warming World – The Transition of Higher Education Institutions to a Low Carbon Economy. River Publishers.

Redshaw, S. (2012). *In the Company of Cars: Driving as a Social and Cultural Practice.* Burlington, VT: Ashgate Publishing, Ltd.

Revi, A. (2008). Climate change risk: an adaptation and mitigation agenda for Indian cities. *Environ. Urbanization* 20:207. doi: 10.1177/0956247808089157

RTO (2015). *Vehicle count crosses 30 Lakh mark in Pune.* Sakal Times. Available at: http://sakaaltimes.com/NewsDetails.aspx?NewsId=52890 34236707686455&SectionId=4924098573178130559&SectionName= Top%20Stories&NewsTitle=Vehicle%20count%20crosses%2030L%20 mark%20in%20Pune [accessed April 15, 2016].

Saveyn, B., Paroussos, L., and Ciscar, J. C. (2012). Economic analysis of a low carbon path to 2050: a case for China, India and Japan. *Energy Econ.* 34, S451–S458.

Schade, J., and Schlag, B. (2003). Acceptability of urban transport pricing strategies. *Transp. Res. F* 6, 45–61.

Schipper, L., Fabian, H., and Leather, J. (2009). Transport and carbon dioxide emissions: forecasts, options analysis, and evaluation. *ADB Sustainable Development Working Paper Series Evaluation*, Manila.

Smith, A., Vo, J. P., and Grin, J. (2010). Innovation studies and sustainability transitions: the allure of the multi-level perspective and its challenges. *Res. Policy* 39, 435–448. doi: 10.1016/j.respol.2010.01.023

Tabachnick, B. G., and Fidell, L. S. (2007). *Using Multivariate Statistics,* 5th Edn. Boston, MA: Allyn & Bacon.

UNFCCC (2015). Available at: http://unfccc.int/meetings/paris_nov_2015/ meeting/8926.php [accessed June 22, 2017].

Vaidyanathan, V., King, R. A., and de Jong, M. (2013). Understanding urban transportation in India as a polycentric system. *Policy Society* 32, 175–185. doi: 10.1016/j.polsoc.2013.05.005

Wang, H., Fang, K., and Shi, Y. (2011). Benefit-cost analysis with local residents' stated preference information: a study of non-motorized transport investments in Pune, India. *J. Benefit Cost Anal.* 2, 1–37.

Wells, P., and Beynon, M. J. (2011). Corruption, automobility cltures, and road traffic deaths: the perfect storm in rapidly motorizing countries? *Environ. Plan. A* 43, 2492–2503.

Zaidel, D. M. (1992). A modeling perspective on the culture of driving. *Accid. Anal. Prev.* 24, 585–597.

Zimmermann, A., Lorenz, A., and Oppermann, R. (2007). "An operational definition of context," in *Proceedings of the International and Interdisciplinary Conference on Modeling and Using Context*, 558–571 (Berlin: Springer).

9

Living Labs for Education for Sustainable Development in the Context of Higher Education: Identifying Triggers and Drivers of Development in the Portuguese Universities

Arminda do Paço[1] and Ulisses M. Azeiteiro[2]

[1]Department of Business and Economics and NECE,
University of Beira Interior, Covilhã, Portugal
[2]Department of Biology and Centre for Environmental
and Marine Studies (CESAM), University of Aveiro,
Aveiro, Portugal

Abstract

Higher Education Institutions (HEIs) perform a relevant role achieving the challenges for sustainability, as well as addressing opportunities for improving academic research, scientific knowledge, and education and training for sustainable development (SD). Several publics are expecting HEI to be sustainable organisations and impact on societal change. As such, living labs for education for SD and co-production are emerging strategies for universities to address sustainability challenges. This chapter provides an overview on the panorama of living labs in HEI. Authors present some successful cases, two from Europe (UK) and three from North America (USA and Canada). For the Portuguese HEI are identified the barriers, challenges and obstacles to instrument sustainable initiatives, as is the case of the living labs implementation. Proposals for embedding sustainability in organisational culture and practice are discussed.

9.1 Introduction

Higher Education Institutions (HEIs) play a crucial role meeting the challenges for sustainability and addressing innovative opportunities for increase scientific knowledge, develop research capacity and knowledge transfer, training and teaching for sustainable development (SD) (Aleixo et al., 2017a,b, in press) and progress has occurred in the last years integrating sustainability in the HEIs (Ferrer-Balas et al., 2008; Shiel and Williams, 2015).

An increasing number of stakeholders expect HEI to be sustainable organisations (Aleixo et al., 2017a,b, in press) and influence societal change (Marcus et al., 2015). The involvement of all the participants in the concept of sustainability is a relevant driver, reason why advances can only be made along the path towards sustainability with the engagement of HEI leaders, faculty staff, students, and external entities (Aleixo et al., 2017a,b, in press).

Living labs and co-production are emerging strategies for universities to address sustainability challenges. This way, living labs can put staff, researchers, students, and external stakeholders (e.g., NGOs, companies, and environmental consultants), together to co-produce knowledge about new sustainability technologies and services in real-world settings. According to Evans et al. (2015) living labs are a way of experimental governance, whereby several stakeholders can develop and test for instance new technologies and lifestyles to address the challenges of climate change and urban sustainability. The process involves the "simulation" of the experiment in a real-world context, which is then monitored and controlled. This approach can be used in several situations and contexts but it is becoming increasingly popular in universities, since their campuses offer all the conditions in which is possible to conduct the applied research. Living lab makes use of a local community to serve as a real infrastructure to develop and test a concept (Koo et al., 2014).

Despite the importance of living labs to the higher education sector, there is a lack of academic research on this topic. As is stated by van Geenhuizen (2013), how living labs perform in reality remains to be documented, particularly the functioning of the networks as well as the models of governance.

This chapter provides a contextualisation and an overview on the literature related to living labs for education for SD in HEI. Authors present some successful cases, two from Europe (UK) and three from North America (USA and Canada). This choice was based in a documental analysis (mainly reports) and web search that returned these cases as the ones frequently designated as successful examples. For the Portuguese HEI, a country in which there is a clear gap in the topic of living labs in the scope of education for SD, the barriers, challenges, and obstacles to instrument sustainable initiatives, are identified. It is expected that this analysis could help HEI to effectively implement living labs for SD. Proposals for embedding sustainability in organisational culture and practice will be presented and discussed.

9.2 Living Labs for Education for Sustainable Development

The concept of living lab is a broader concept, which can be used and applied not only in HEI but also into several contexts, as can be inferred from the following definition: "... the development of new products, systems, services, and processes, employing working methods to integrate people into the entire development process as users and co-creators, to explore, examine, experiment, test and evaluate new ideas, scenarios, processes, systems, concepts and creative solutions in complex and real contexts" (JPI Urban Europe, 2013). Evans et al. (2015) adds that living labs address practical problems related to infrastructure, built design, low carbon technologies and urban problems, through collaborative experiments that integrate both users and stakeholders. The model underlying is based in a real place in which is possible to set problems with several publics. Living labs can frame co-production processes because putting together users and providers will allow to discuss iteratively and improve holistic solutions to face sustainability challenges taking advantage of the process of experimenting and learning. Thus, living lab projects should aim[1] to:

1. Solve and explain a real life problem;
2. Be based on a partnership among relevant stakeholders (often from diverse sectors);
3. Use existing and new quantitative and qualitative data;
4. Trial and test ideas in real life situations;
5. Share/disseminate data and analysis generated openly.

_____

[1]http://www.ed.ac.uk/about/sustainability/themes/research-teaching/the-university-as-a-living-lab [accessed on March 17, 2017].

According to van Geenhuizen (2013), the concept of living labs is attributed to William Mitchell, at the Massachusetts Institute of Technology in 2003, who proposed to move various types of research from laboratories to *in vivo* environments monitoring users' responses and interactions with innovations. In this sense, the aim was to speed up the development of the innovation and make it more effective since this way it was possible the matching from the first moment of the creation using delimited real contexts (e.g., hospital, city, university campus, etc.). Thus, living labs are places and open networks to test solutions for problems of clients and communities that can be adapted to several contexts.

The European Commission defends and supports actively the concept of living labs, seeing these as a way to increase the level of innovativeness of European countries. Thus, a pan-European network of living labs (European Network of Living Labs) was launched (European Commission, 2010), first focusing only the introduction of new ICT tools, but later extending to other fields (sustainable energy, health care, and safety).

According to Graczyk (2015), the definitions of living labs vary a lot depending on the environment that the concept is embedded in and the desired outcome. Moreover, there is a diversity of tools and methodologies that could be chosen to deliver these projects, including field experiments or context mapping. Although there are discrepancies of the definition in the literature, there has been an agreement on the common purpose of the living labs being generally described as an environment in which new solutions are evaluated or validated with all relevant stakeholders to create innovation (Dell'Era and Landoni, 2014; Budweg et al., 2011).

Graczyk (2015) points a reason for such divergence on the definition. It can be justified with the fact that there are many different types of living labs environments, such as: research living labs focusing on performing research on the innovation process; corporate living labs focusing on co-creating innovations with the stakeholders (e.g., citizens); organisational living labs where the members of an certain entity co-develop innovations; and intermediary living labs in which different partners are invited to collaboratively innovate in a neutral action field.

Living labs can contribute to accelerating knowledge valorisation. Spend high amounts of time and money searching for business partners and for bringing knowledge to a market can be avoided or diminished by HEI and companies. In specific, living labs seem to potentiate the relationship between both organisations. This effect can be enlarged if other regions or even countries where involved (van Geenhuizen, 2013).

The living lab in the HEIs scope promotes applied research and education by using the campus to test real-time sustainability solutions, offering opportunities to all stakeholders to participate in the process turning the theory into practice, allowing students to achieve greater engagement with their educational programme (International Alliance of Research Universities, 2014).

By one side, living labs enable applied research and provide real world experience that will prepare students for a competitive job market, and by the other side, living labs for education for SD can catalyse new styles of teaching and learning that contribute to the sustainability of both universities and cities (Evans et al., 2015). The advantages to students involved in the living lab process are several. They assist in the development of infrastructure system renewal to achieve sustainable living conditions (Koo et al., 2014).

van Geenhuizen (2013) refers some critical factors that can contribute to the successful creation and management of living labs, as the involvement of the users allowing for co-creation and testing; the balanced composition of participants; the use of adequate functioning models; the dealing with potentially troubling legal issues; and the team. Usually, a team for the living lab project consists of many diversified and interdisciplinary members, including students, faculty staff, community leaders, users, and firms (Koo et al., 2014). Regarding the obstacles, usually the main relevant are related with the resources time and money (Graczyk, 2015).

9.3 Sustainability on Campus: Some Successful Cases

In this section some successful cases, two in Europe (in United Kingdom, the Manchester, and the Edinburg universities) and three in North American (in United States of America, the Indianapolis and Yale, and in Canada the British Columbia universities) are presented. As is referred by van Geenhuizen (2013), the living labs as a concept used by universities, has been present for a long time, especially in USA and Canada, and is now making its way to the Europa and especially in UK. Thus, there are some very successful cases of universities (and colleges) that have been implementing living labs for education for SD within their campuses, benefiting the students, staff, and community. Here, we will present the case of Manchester, Edinburg, Indiana, Yale, and Columbia universities.

In the University of Manchester, in UK, the sustainability is seen as part of its broader social responsibility agenda, which was institutionalised as a top-level goal. This institution has been applied the teaching and learning named the "Manchester Method", which implies working with external stakeholders to develop also non-academic skills. Despite the existence of interdisciplinary courses on sustainability, there are other activities, including a community of practice focusing on "Embedding Sustainable Development" (Evans et al., 2015).

A huge investment programme of building and refurbishments was carried out in the last years, and in response to this opportunity, the University Living Lab initiative was launched in 2012 to transform the campus into a place for applied teaching and research around sustainability, providing a systematic framework for students and academics to engage with the opportunities to work with the Estates staff and their environmental consultants on applied sustainability challenges (University of Manchester, 2015). One of the aims of this structure is acting as a supporting tool for meeting the 40% carbon reduction target set by the university (Graczyk, 2015).

The living lab team communicates with course leaders to understand the subject focus, research requirements and deadlines of specific programmes. This has resulted in a huge list of projects focused on ongoing refurbishment projects, new campus developments, procurement and supply chain innovation, etc. Thus, on the site is possible to find people, courses and projects relating to specific research areas. Some of the most successful projects are: the "Sackville Street Student Lab" that is an ambitious attempt to hand over control of the lab to students for experimentation with electrics, lighting, ventilation, and ambient temperature; the "Manchester Cycling Lab" that intends to turn Manchester into a real-life laboratory for the study of cycling, involving working in collaboration with Manchester City Council, Transport for Greater Manchester (TfGM) and local businesses to identify the gaps[2]; between other projects.

It is possible to notice the following aspects: (i) living lab research projects require particular motivation from the students, as the research will feed directly into ongoing work on campus and involves a range of external stakeholders. However, students see this as a positive challenge and recognise the value of this experience for their future careers; (ii) further, projects require a commitment from academics to work in other ways, sometimes "outside of the box"; and (iii) the inclusion of living lab student projects into the curriculum should not be seen as a disjointed sustainability initiative but is part of a wider drive towards applied learning and employability skills (Evans et al., 2015).

---

[2]http://universitylivinglab.org/projects [accessed on March 09, 2017].

The University of Edinburg is itself a living lab since it involves all professors and researchers, non-academic staff and students to solve social responsibility and sustainability issues. Its living lab is a citywide collaboration engaging the municipality and the university, bringing the academy, the public sector, the companies, and the non-profit organisations together in order to work in a holistic way with citizens in co-designing, testing and implementing new services, processes and products that generate social, environmental, and economic value. This iterative process of "experimenting/refining/redefining" is based in a participatory methodological research with the end users through small-scale experiments, which contributes to developing a better understanding of the problems and needs.

The focus of the Edinburg living lab are the (i) mobility (to explore the interactions between factors as transport, environmental and waste management, development planning, city liveability, education, and well-being); (ii) energy (to analyse how people interact with energy in their daily lives and provide better and more environmentally friendly solutions); and (iii) learning by developing (to use a pedagogical model that builds the learning experience around real-life problems, allowing students to engage with problems external to the academy, and testing their own abilities)[3].

Over time, the Edinburg living lab has been working into several projects. For instance, in the scope of the energy management in public sector buildings, the "Enhance", a collaborative research project on energy, aims to use data analytics to provide feedback to building users to understand and reduce energy demand. The interdisciplinary team working in this project is taking a holistic approach, researching the effects of organisational behaviour on energy use. In practice, several solutions as the installation of additional sensors in buildings can be better planned. Other interesting project is the "Wrangling Edinburgh Bike Counter Data". The Edinburgh Council has installed automatic bike counters across the city, most of them on the network of off-road bike paths. There were a need to get hold of the data collected by the counters using different sources of cycling data. One of the objectives was getting access to previously unpublished data collected and see it available in the Council's open data portal (what really happen).

The Indiana University Purdue University Indianapolis (IUPUI), in USA, is committed to providing educational opportunities that transform the lives of its students and community, being a national leader in urban sustainability while supporting student success and the well-being of citizens. As such, the

---

[3]http://edinburghlivinglab.org/ [accessed on March 17, 2017].

university is very involved in the development of its living lab. One of the successful examples is presented by Koo et al. (2014) and involves a local neighbourhood located close to the IUPUI campus, which has experienced a socio-economic decline and its underground infrastructure system is really needing of restoration. This living lab project was designated as the "Riverside Watershed Environmental Living Lab for Sustainability" (RWELLS) and aimed to: involve the residents of the community making them aware of the importance of water conservation; introduce smart water practices; create solutions with the industry partners so that can be implemented and replicated to other communities; and support students' research and initiatives. The project is focused on underground infrastructure assessment, development of new solutions, and promotion of entrepreneurship opportunities within the community.

The student team learned how to reconcile the challenges of working within a multi-disciplinary working environment as it attempted to solve technical problems for the community, all the while remaining in-budget and sensitive to the socio-economic needs of the community. RWELLS demonstrated that a local community can serve as a living lab to establish an effective extracurricular project. The project provided a practical student experience of professional services. The most fundamental aspect of this extracurricular program is that industry and university must collaborate effectively and seek mutual benefits from the students and society (Koo et al., 2014).

The University of Yale commitment with sustainability is well settled in its "Yale Sustainability Plan 2025". This plan is based on a vision where sustainability is completely integrated into the scholarship and operations of the institution, contributing to its social, environmental, and financial excellence, and positioning at national and global level. Its purpose is really aspiring being founded in nine ambitions (empowerment, leadership, technology, materials, mobility, built environment, stewardship, climate action and health, and well-being). These are detailed by 20 objectives and 38 goals. Each goal is sustained by several strategies and key indicators (University of Yale, 2016).

The living lab of the University of Yale[4] is supported by the Office of Sustainability, which encourages the use of the institution and the city of New Haven as living laboratories for multidisciplinary sustainability

---

[4]http://sustainability.yale.edu/research-education/campus-living-labYale [accessed on March 15, 2017].

research, focusing aspects related to ecological, human health, and socio-economic issues. As such, several projects have been implemented. For example, the "Yale Carbon Charge Project" is testing the effectiveness and feasibility of carbon pricing on Yale's campus. Using the university as a living laboratory for applied research (involving 20 university buildings during 6 months), the project aims to inform energy policy, climate change mitigation, and environmental economics by testing multiple models of carbon pricing. At the end, the idea is to diminish campus energy costs and greenhouse gas emissions by using financial incentives to boost behaviour and decisions aligned with the principles of a low-carbon economy. Other example is the "Compost Tea Study", a collaborative initiative between students and faculty from the School of Forestry and Environmental Studies, Yale Grounds and Maintenance, and the Office of Sustainability. The joint pilot project will monitor four different treatment protocols at eight test sites across campus. It intends to compare the effects of a compost tea amendment, made from Yale's composted food waste, versus current treatment methods on the ecology of Yale's soils. The general aim is to develop a pilot project testing sustainable landscape management practices on campus and study the use of compost tea versus synthetic fertiliser and herbicide.

In University of British Columbia (UBC) in Canada sustainability "is not just a word to define". There is a collective effort in education, research, partnerships, and operations, advancing sustainability on the campus and beyond. The university has described four student sustainability attributes—Holistic Systems Thinking, Sustainability Knowledge, Awareness and Integration, and Acting for Positive Change—to help guide academic units to develop sustainability learning paths (Marcus et al., 2015). This institution has been recognised as one of the institutions that has managed to introduce the most successful living lab concept. It was created in 2010 aiming to integrate operational and academic sustainability across the university and to make its campus available as a kind of societal test bed. It operates integrating partnerships between university and the private, public, and NGO sectors and transferring knowledge to the wider community. The "20 Year Sustainability Strategy" was approved in 2014, covering a wide range of activities including an higher focus on developing research within and outside the university and involving strategic partnerships with industry and government; a new focus on university operations and infrastructure through the vision of the living lab; and regarding teaching and learning, an improved institutional commitment to embed sustainability learning across all undergraduate teaching programmes by 2035 (University of British Columbia, 2014).

The living labs projects very often have a strong business component. Some of them are based on large scale developments, such as the creation of a biomass gasification co-generation plant, constructing outstanding buildings that score the highest standards of sustainability certifications, or a optimisation programme in partnership with the local energy company where they are going to re-commission and recalibrate energy and water systems in a set of buildings (Graczyk, 2015).

9.4 The Portuguese Case: Barriers and Challenges

Given the examples above exposing some of the best practices around the world, is now time to make some reflections about the Portuguese HEI. Aleixo et al. (2016) investigated how SD has been incorporated into Portuguese HEI reviewing the institutional websites for the Public HEI in Portugal and conclude that SD practices are actively communicated by the majority of HEI. Recently, the same authors analysed the implementation of SD practices in the Portuguese HEI through the application of a questionnaire to institutional leaders (rectors, presidents, directors of faculties, departments and schools of Portuguese universities, and polytechnics) finding that Portuguese HEIs give more emphasis to the economic and social dimensions being the environmental dimension the least developed (Aleixo et al., in press). Following this description in this section, some barriers, challenges, and obstacles implement sustainable initiatives, as is the case of the living labs for education for SD, are identified.

The implementation of living labs in Portugal started in the 90's and since then it has played a central role in the socio-economic development and becoming a reference in the socio-economic development of the country (de Oliveira and Amaral de Brito, 2013). Some Portuguese Universities engage in this concept, structures, and activities (e.g., University of Minho, through its School of Engineering and its research centres lead a regional initiative; University of Aveiro through the partnership with the SJM-ILL—S. João da Madeira Industrial living lab; and other living labs gather between their associates HEI like the Intelligent Sensing and Smart Services Living Lab ISaLL in Coimbra and the Living Lab of Cova da Beira with UBI) being active partners in the synergies established.

High educational institutions conduct research within the living lab partnership which actively disseminate and participate in the communication strategies and policies. There is a clear interface and knowledge transfer intense activity within this partnerships. However, there are no living labs within the Portuguese HEI as emerging strategies for universities to address sustainability challenges.

Velazquez et al. (2006, p. 812) defines Sustainable Higher Education Institutions (SHEIs) as "a HEI (...) that addresses, involves and promotes, on a regional or a global level, the minimisation of negative environmental, economic, societal, and health effects generated in the use of their resources in order to fulfil its functions of teaching, research, outreach and partnership, and stewardship in ways to help society make the transition to sustainable life-styles". More recently Aleixo et al. (2017b, p. 2) reinforces that HEIs should consider the issues of SD "through all structural and organisational dimensions, infrastructure and energy related aspects, the efficient use of resources, by on-going strategic actions in education, research, knowledge transfer and with stakeholders (partnerships and community)".

High educational institutions must be better prepared to intervene in the promotion of sustainability, namely leading by example (Amaral et al., 2015, p. 156) and an increasing number of stakeholders expect them to be sustainable organisations (Aleixo et al., 2017a). Some difficulties prevail. The reported difficulties for the HEI, namely the ones recently divulged by Aleixo et al. (2017b), related to financial constraints and stakeholders perceptions, difficult the appearance and consolidation of living labs within the Portuguese HEI as emerging strategies for universities to address sustainability challenges.

Some of the barriers to sustainability (resistance to change, cultural, and behavioural change and low commitment and lack of engagement) can be seen as general barriers related to the institutional culture. Lozano et al. (2013) defend that SD should be imbedded in the mission of the university and thus, implementing SD through campus experiences, by incorporating SD into the daily activities in the university experiences and "educating-the-educators" is crucial. Recently Ávila et al. (2017) published a very comprehensive study about barriers to innovation and sustainability at universities were obstacles are summarised and the relations between innovation and sustainability are discussed. Ávila et al. (2017) reinforce the need for greater support from university administrations and management as already concluded by Aleixo et al. (2017a,b, in press) from the Portuguese reality.

The challenge in HEI is the same as for living labs within the Portuguese HEI that is to promote the change. This change promoters identified by Aleixo et al. (2017b) for the HEIs (e.g., conceptual and organisational change, identifying new sources of financing, more flexible organisational forms against the rigid organisational structure which is characterised by many hierarchical levels, more comprehensive mission statements, more tailored educational offers, lifelong learning and commitment to internationalisation, and more strategic human resource management fostering commitment, engagement, awareness, interest, and involvement of most stakeholders) are the same obstacles for the consolidation of living labs within the Portuguese HEIs.

This theme has been researched for some time (e.g., Lozano, 2006; Adams, 2013; Barth, 2013) and more recently Ávila et al. (2017) and Aleixo et al. (2017b), that review and address these challenges in the Portuguese HEIs, discuss measures to overcome the barriers and obstacles. These measures are proactive leadership and commitment of institutional leaders, support of senior management, consistent communication and flexible organisational structures, the inclusion of sustainability in the HEIs strategies, multidisciplinarity in research and courses (integrating sustainability issues in education, research, operations and outreach, and collaboration with the community), engagement of students (to empower students with skills to address the problems of society for future well-being) and staff and initiatives that develop engagement in sustainability practices and support systems. These are the kind of recommendations and measures that are replicable for the implementation and consolidation of living labs within the Portuguese HEIs.

Regarding living labs within the Portuguese HEI as emerging strategies for universities to address sustainability challenges, and coting Lozano (2006) about HEI having a growing number of SD practices, within and outside the institutions, there is a need for a systematic and holistic perspective/vision. Integration of SD into HEI systems (integrated into the whole system including curricula, research, campus operations, community outreach and partnerships, and assessment and reporting) is thus defended by several authors (see Aleixo et al., 2017a,b, in press, 2017). Leal Filho et al. (2015) add that integrative efforts are not only expected to last longer but also yield tangible benefits to the community. Thus, an institutional predisposition to fortify the internal capacity in the field of SD as a whole, and to establish and develop a framework which encourages the research and practical work performed by members of staff, is needed.

9.5 Conclusion

There are several successful cases of HEI that have implemented living labs within their campuses, which have in turn benefited the students, the staff, and the community surrounding. The solutions founded had very often resulted in a reduction of the carbon footprint, in opportunities to students learn by practicing, in enhancing social sustainability performance; and in using institutional resources efficiently.

As saw in the examples presented, the key strengths of the living lab for education for SD approach are that it provides a systematic approach to facilitate student (and academic) engagement with applied sustainability issues. There is a co-production through consultation between non-academic and academic stakeholders within a certain institutional and geographical context. Working in this way maximises the benefit of knowledge produced to non-academic stakeholders and motivates the students since they have a contact with the real-word.

As is evidenced for instance in the experience of Manchester, focusing on specific problems or infrastructures is effective as it allows the creation of a community of interest, while working with external stakeholders shows clear pathways to real-world impact, generating more holistic solutions.

In terms of policy recommendation, at least in Portugal there is a need to better clarify the concept of living labs; it is necessary to identify which network structures and forms of governance are the best. The discussion about these structures started recently with the Portuguese Agency for Innovation promoting meetings with several partners, including the technological centres and the applied research laboratories, to debate the new agendas for research and innovation, and the partnerships between companies and science, including the development of collaborative labs in Portugal. Additionally, the Portuguese Foundation for Science and Technology made available a project for the regulation for the attribution of the title of Collaborative Lab (CoLAB)[5]. This document clarifies that the Collaborative Laboratories are associations of research units, associated laboratories, HEI, technological centres, companies, business associations, and other relevant partners (e.g., laboratories of the State, hospitals, museums, etc.). The main aim is to define and implement research and innovation agendas oriented to the creation of economic and social value, including the internationalisation process of the national scientific and technological capacity in relevant areas of intervention and the stimulation of scientific employment. This is aligned with the main purpose of collaborative living lab projects; this is, to provide

---

[5]https://www.fct.pt/noticias/index.phtml.pt?id=216&/2017/2/Projeto_de_Regulamento_de_Atribui%C3%A7%C3%A3o_do_T%C3%ADtulo_de_Laborat%C3%B3rio_Colaborativo_(CoLAB)_dispon%C3%ADvel_no_site_da_FCT [accessed March 16, 17].

responses and guidance for operations and professional services staff, real-life learning opportunities for students, and opportunities for research impact for academics.

At the same time, local/regional policy makers, and universities should put more energy in developing the necessary infrastructures in the region, as are the case of the living labs promoting the proximity between all stakeholders, including the community.

Acknowledgements

FCT (the Portuguese Foundation for Science and Technology) and POCH (Programa Operacional do Capital Humano – Operational Programme for Human Capital).

References

Adams, C. A. (2013). Sustainability reporting and performance management in universities. *Sustain. Account. Manage. Policy J.* 4, 384–392.

Aleixo, A. M., Azeiteiro, U., and Leal, S. (2016). "Toward sustainability through higher education: sustainable development incorporation into Portuguese Higher Education Institutions," in *Challenges in Higher Education for Sustainability: Management and Industrial Engineering*, eds J. P. Davim and W. Leal Filho (Cham: Springer), 159–187.

Aleixo, A. M., Azeiteiro, U. M., and Leal, S. (in press). The implementation of sustainability practices in portuguese higher education institutions. *Int. J. Sustain. High. Educ.*

Aleixo, A. M., Azeiteiro, U. and Leal, S. (2017a). "UN decade of education for sustainable development: perceptions of higher education institution's stakeholders," in *Handbook of Theory and Practice of Sustainable Development in Higher Education: World Sustainable Development Series,* Vol. 4, eds W. Leal Filho, U. M. Azeiteiro, F. P. Alves, and P. Molltan-Hill (Berlin: Springer), 417–428.

Aleixo, A. M., Leal, S., and Azeiteiro, U. M. (2017b). *Conceptualization of Sustainable Higher Education Institutions, Roles, Barriers, and Challenges for Sustainability: An Exploratory Study in Portugal.* Available at: http://dx.doi: 10.1016/j.jclepro.2016.11.010

Amaral, L. P., Martins, N., and Gouveia, J. B. (2015). Quest for a sustainable university: a review. *Int. J. Sustain. High. Educ.* 16, 155–172.

Avila, L. V., Leal Filho, W., Brandli, L., MacGregor, C., Molthan-Hill, P., Ozuyar, P. G., et al. (2017). Barriers to innovation and sustainability at universities around the World. *J. Clean. Prod.* 164, 1268–1278.

Barth, M. (2013). Many roads lead to sustainability: a process-oriented analysis of change in higher education. *Int. J. Sustain. High. Educ.* 14, 160–175.

Budweg, S., Schaffers, H., Ruland, R., Kristensen, K., and Prinz, W. (2011). Enhancing collaboration in communities of professionals using a Living Lab approach. *Prod. Plan. Control* 22, 594–609.

de Oliveira, A., and Amaral de Brito, D. (2013). Living Labs: a experiência portuguesa. *Rev. Iberoam. Cienc. Tecnol. Soc.* 8, 201–229.

Dell'Era, C., and Landoni, P. (2014). Living Lab: a methodology between user centred design and participatory design. *Creat. Innov. Manag.* 23, 21–18.

European Commission (2010). *Advancing and Applying Living Lab Methodologies: An Update on Living Labs for User-driven Open Innovation in the ICT Domain.* Luxembourg: Publications Office of the European Union.

Evans, J., Jones, R., Karvonen, A., Millard, L. and Wendler, J. (2015). Living labs and co-production: university campuses as platforms for sustainability science. *Curr. Opin. Environ. Sustain.* 16, 1–6.

Ferrer-Balas, D., Adachi, J., Banas, S., Davidson, C. I., Hoshikoshi, A., Mishra, A., et al. (2008). An international comparative analysis of sustainability transformation across seven universities. *Int. J. Sustain. High. Educ.* 9, 295–316.

Graczyk, P. (2015). *Embedding a Living Lab Approach at the University of Edinburgh.* Edinburgh: University of Edinburg.

International Alliance of Research Universities (2014). *Green Guide for Universities. IARU Pathways towards Sustainability.* International Alliance of Research Universities (IARU).

JPI Urban Europe (2013). *Urban Europe: Creating Attractive, Sustainable and Economically Viable Urban Areas.* Available at: http://jpi-urbaneurope.eu/ [accessed December 18, 14].

Koo, D., White, J. W., and Ray, V. M. (2014). "Development of effective extracurricular construction technology education programs for university and industry collaborations," in *Proceedings of the 121st ASEE Annual Conference and Exposition*, (Washington, DC: American Society for Engineering Education).

Leal Filho, W. (2000). Dealing with misconceptions on the concept of sustainability. *Int. J. Sustain. High. Educ.* 1, 9–19.

Leal Filho, W., Shiel, C. And Paço, A. (2015). Integrative approaches to environmental sustainability at universities: an overview of challenges and priorities. *J. Integr. Environ. Sci.* 12, 1–14.

Lozano, R. (2006). Incorporation and institutionalization of SD into universities: breaking throughbarriers to change. *J. Clean. Prod.* 14, 787–796.

Lozano, R., Lukman, R., Lozano, F. J., Huisingh, D., and Lambrechts, W. (2013). Declarations for sustainability in higher education: becoming better leaders, through addressing the university system. *J. Clean. Prod.* 48, 10–19.

Marcus, J., Coops, N., Ellis, S., and Robinson, J. (2015). Embedding sustainability learning pathways across the university. *Curr. Opin. Environ. Sustain.* 16, 7–13.

Shiel, C., and Williams, A. (2015). "Working together, driven apart: reflecting on a joint endeavour to address sustainable development within a university," in *Proceedings of the Integrative Approaches to Sustainable Development at University Level*, (Cham: Springer International Publishing), 425–447.

University of British Columbia (2014). *20-Year Sustainability Strategy for the University of British Columbia, Vancouver Campus*. Available at: http://sustain.ubc.ca/sites/sustain.ubc.ca/files/uploads/CampusSustainability/CS_PDFs/PlansReports/Plans/20-Year-Sustainability-Strategy-UBC.pdf [accessed 09.03.17].

University of Manchester (2015). *University of Manchester Environmental Sustainability*. Available at: http://www.sustainability.manchester.ac.uk [accessed March 9, 17].

University of Yale (2016). *Yale Sustainability Plan 2025*. Available at: http://sustainability.yale.edu/sites/default/files/sustainability_plan_2025.pdf [accessed March 17, 2017].

van Geenhuizen, M. (2013). From ivory tower to living lab: accelerating the use of university knowledge. *Environ. Plan. C* 31, 1115–1132.

Velazquez, L., Munguia, N., Platt, A., and Taddei, J. (2006). Sustainable university: what can be the matter? *J. Clean. Prod.* 14, 810–819.

Index

About the Editors

Ulisses M. Azeiteiro is a Senior Professor (Associate Professor with Habilitation and Tenure), Coordinator of the Climate Change and Biodiversity Assets Unit from the Biology Department and Integrated Member/Senior Researcher of the Centre for Environmental and Marine Studies (CESAM) at University of Aveiro in Portugal. His main interests are the Impacts of Climate Change and Adaptation to Climate Change in the Context of Sustainable Development (Social and Environmental Sustainability and Climate Change). Research Coordinator since 2008 he was a member of several evaluation panels from research projects and grants (International, European and National—Portuguese and outside Portugal), member of the organizing and scientific committees of more than 100 international and national congresses, supervised over 50 postgraduate students (M.Sc. and Ph.D. students). Professor Ulisses M. Azeiteiro has written, co-written, edited or co-edited more than 250 publications, including books, book chapters and papers in refereed journals.

Walter Leal Filho is a Professor of Environment and Technology at Manchester Metropolitan University (UK) and Hamburg University of Applied Sciences (Germany), where he heads the Research and Transfer Centre "Sustainable Development and Climate Change Management" and the World Sustainable Development Research and Transfer Centre.

João Paulo Davim received the Ph.D. degree in Mechanical Engineering in 1997, the M.Sc. degree in Mechanical Engineering (materials and manufacturing processes) in 1991, the Licentiate degree (5 years) in Mechanical Engineering in 1986, from the University of Porto (FEUP), the Aggregate title (Full Habilitation) from the University of Coimbra in 2005 and the D.Sc. from London Metropolitan University in 2013. He is Eur Ing by FEANI-Brussels and Senior Chartered Engineer by the Portuguese Institution of Engineers with a MBA and Specialist title in Engineering and Industrial Management. Currently, he is Professor at the Department of Mechanical

Engineering of the University of Aveiro, Portugal. He has more than 30 years of teaching and research experience in Manufacturing, Materials and Mechanical Engineering with special emphasis in Machining & Tribology. He has also interest in Management & Industrial Engineering and Higher Education for Sustainability & Engineering Education.

Lightning Source UK Ltd.
Milton Keynes UK
UKHW021201130219
337258UK00001B/3/P